KARST

An Introduction to Systematic Geomorphology
VOLUME SEVEN

Karst

J. N. JENNINGS

Professorial Fellow in Geomorphology,
The Australian National University

THE M.I.T. PRESS
Cambridge, Massachusetts and London, England

ISBN 0 262 10011 8 (hardcover)

Library of congress catalog card number: 45-169945

PRINTED IN AUSTRALIA

INTRODUCTION TO THE SERIES

This series is conceived as a systematic geomorphology at university level. It will have a role also in high school education and it is hoped the books will appeal as well to many in the community at large who find an interest in the why and wherefore of the natural scenery around them.

The point of view adopted by the authors is that the central themes of geomorphology are the characterisation, origin, and evolution of landforms. The study of processes that make landscapes is properly a part of geomorphology, but within the present framework process will be dealt with only in so far as it elucidates the nature and history of the landforms under discussion. Certain other fields such as submarine geomorphology and a survey of general principles and methods are also not covered in the volumes as yet planned. Some knowledge of the elements of geology is presumed.

Four volumes will approach landforms as parts of systems in which the interacting processes are almost completely motored by solar energy. In humid climates (Volume One) rivers dominate the systems. Fluvial action, operating differently in some ways, is largely responsible for the landscapes of deserts and savannas also (Volume Two), though winds can become preponderant in some deserts. In cold climates, snow, glacier ice, and ground ice come to the fore in morphogenesis (Volume Three). On coasts (Volume Four) waves, currents, and wind are the prime agents in the complex of processes fashioning the edge of the land.

Three further volumes will consider the parts played passively by the attributes of the earth's crust and actively by processes deriving energy from its interior. Under structural landforms (Volume Five), features immediately consequent on earth movements and those resulting from tectonic and lithologic guidance of denudation are considered. Landforms directly the product of volcanic activity and those created by erosion working on volcanic materials are sufficiently distinctive to warrant separate treatment (Volume Six). Though karst is undoubtedly delimited lithologically, it is fashioned by a special combination of processes centred on solution so that the seventh volume partakes also of the character of the first group of volumes.

J. N. Jennings
General Editor

PREFACE

This small book perforce omits many kinds of karst features and even more of the ideas karst has prompted. Because of personal experience, Australia, New Zealand, New Guinea, and Malaysia loom larger than their place in karst literature might direct but an outlook other than European or North American is part of the object of this series. I have preferred enlarging on the elements to pursuing the complex, and regional 'personality' has been omitted altogether. Recent morphometric experimentation and process study have been given more space than product from these approaches probably warrants at the moment, yet the growing points deserve some favour. My hope is that this restricted selection of examples and ideas will entice readers to more thorough books on the subject and above all to karst itself where one cannot fail to be seized by the endless round of rock and water, rock consumed by water, water creating rock.

The book has benefited greatly from the suggestions and criticism of my friends Mr C. D. Ollier and Dr P. W. Williams to whom I am particularly grateful. However little it will satisfy him, Chapter VIII is in debt to Mr R. M. Frank without his formal help. I have consulted Dr K. Crook on limestone classification, Dr R. Rosich on the chemistry of limestone solution, and Mr L. Milton on piping. I am indebted to them all for aid, though the inadequacies of the book remain mine. Mr K. Fitchett helped in innumerable ways in both field and laboratory. Mr 'Ian' Heyward, who drew the figures, was both patient and constructive. I thank both of them and all those who contributed photographs and on whose work the line figures are based. Last but not least I must acknowledge my memory-filling debt to the goodly company from several countries, who have shown me caves or gone caving with me.

Canberra   J.N.J.
1970

CONTENTS

FIGURES

PLATES

TABLES

I

THE NATURE OF KARST AND ITS STUDY

Derived from the Slav word *Krš*, meaning crag or stone and also the geographical name of limestone country in western Slovenia, the German word *Karst* has passed into international usage in an extended and imprecise way. It is therefore necessary to explain how the term will be used in this book and to comment on other usage. This discussion of the meaning of 'karst' will also serve to introduce its study.

Karst signifies *terrain with distinctive characteristics of relief and drainage arising primarily from a higher degree of rock solubility in natural waters than is found elsewhere.* The word is also used adjectivally to refer to rock, water, streams, caves and other features making up such landscape.

Karst is frequently marked by intermittent stream flow and by valleys without stream channels—dry valleys—yet it cannot be adequately defined in terms of replacement of surface by underground drainage. Much distinctive tropical karst has predominantly surface drainage; pumice and loess areas are often virtually without it yet fail to develop other attributes common in karst.

Continuous systems of slopes and channels taking water to the sea typical of fluvial relief give way in karst to apparently disorganised, even chaotic relief in which valleys are frequently interrupted and where there are many types of closed depression. Again, however, some kinds of karst lack these characters which conversely are found in country other than karst. In a dry climate, for instance, river action may not be persistent enough to overcome the disruptive effects of earth movements on valley systems or to offset the power of wind to hollow out basins. In dune fields the primary hollows resulting from aeolian deposition may remain

1

closed for a long time, even in humid regimes, because rainfall soaks easily into the sand and streams cannot integrate the fields of hollows into valley networks.

Nor can definition be solely in terms of the morphology of limestone since similar relief and drainage are developed on other rocks. The topography associated with the full range of carbonate rocks is a better basis since dolomite is the next most widespread rock type giving rise to karst. But in addition to carbonate rocks, very soluble evaporites such as halite, gypsum, and anhydrite induce karst, especially in dry climate as with the Castile Formation of northwest Texas. At the opposite extreme in very humid climate some karst characteristics develop in less soluble rocks such as quartz diorite in northern Colombia (Feininger 1969) and eclogite in the Owen Stanley Range of New Guinea.

The definition adopted in this book stresses solubility. However, all rocks are soluble in natural waters to some extent and in karst itself solution need not be the only or even the dominant process, but it does play a significantly more important role here than in other landscapes. Solution operates in more than one way to produce karst but by far the most important effect lies in the enlargement of voids in the rock. The outcome is the steady growth of permeability. The resulting capacity to transmit large amounts of water rapidly is responsible for the development of underground drainage and the disruption of valley systems.

Solution of this localised kind can produce voids which are penetrable so caves are more frequent, more elaborate, and larger in karst than in any other terrain. In all branches of geomorphology there is substantial involvement with subsurface data, for example, with the thickness of superficial deposits or the nature of buried surfaces, but only in karst can one truly speak of 'underground geomorphology' wherein a wide variety of erosional and depositional forms demand survey and explanation. This aspect of karst therefore requires serious consideration in this book though the balance of treatment between external and internal forms must favour the former, whereas in works of speleology the reverse is properly the case.

Precipitation from natural waters is important in several morphogenic systems; witness salt pans in deserts and laterites in seasonally humid tropics. But deposition from solution looms very large in karst where its significance also varies with climate.

Along the coasts of karst country solution and deposition of carbonate make for distinctive assemblages of landforms. This aspect of the subject will not be dealt with here and readers are referred to the book on *Coasts* in this series (Bird 1968).

Apart from coastal karst very many kinds of karst have been recognised, some of which will be discussed later. However, certain distinctions are so general that it will help to touch on them now.

The most common contrast made is that between *bare* and *covered karst* where bedrock is largely exposed to the atmosphere in the former (Pl. 33), modestly or not at all in the latter (Pl. 22). Clearly there will be every stage of transition between the two. The cover may be residual soil and organic matter accumulating in place, or transported unconsolidated deposits of aeolian, glacial, fluvial and other genesis. J. F. Quinlan (unpublished paper) proposes *subsoil* and *mantled karst* to distinguish the two, though it is essential to retain the general term because it is often difficult to discriminate between them and they certainly occur together like a patchwork quilt. Karst features beneath transported materials may be older or younger than their cover whereas those associated with residual soils are syngenetic, developing as insoluble and organic fractions accumulate.

Karst features, which may or may not affect the surface as mentioned above, also develop beneath other bedrock formations (Penck 1924). The term 'covered karst' is not normally employed for this condition, for which the term *subjacent karst* following Martin (1965) is adopted here. The overlying rocks may be other marine beds and the contact may be conformable or unconformable. Alternatively they may be terrestrial outpourings of lava, for example. With conformable relationships there is no doubt that the karst features are younger than the deposition of the overlying strata. This is essentially true, for example, of the karst of the Carboniferous Limestone on the northern flank of the South Wales Coalfield, in which large cave systems have developed under the overlying Millstone Grit. The surface of the Millstone Grit has also acquired a karst landscape as a result (Thomas 1963).

When the strata are unconformable, however, there is always the possibility that some features may be inherited from the time before the overlying rocks were laid down. Surface karst forms may be overlain by later geological formations and caves completely filled by this process. This *buried karst* is also termed *fossil karst* or *palaeokarst*. Quarrying exposed a cave completely filled with Devonian rocks near Devonport, Tasmania (Burns 1964) and by exploring old lead mines Ford and King (1966) have shown that a portion of the large Golconda Cave in the Peak District, England, had been filled with Permo-Triassic mineralised breccia and Tertiary quartz sands. Infilled closed surface depressions may come to light through quarrying and their fillings can range over long periods of geological time (Gilewska 1964; Ford and King 1969). Natural erosion also can expose fossil karst features. This gives rise to *exhumed karst*; it is more likely to happen and to be more readily detected with projecting, rather than recessive forms. In central Manus Island, New Guinea, conekarst (p. 184) is being uncovered by removal of Pliocene lavas (P. W. Williams, pers. comm.). Whether the reef knolls of parts of the Lower Carboniferous karsts of the British Isles belong here is debatable. Most of these knolls represent submarine biohermal masses of limestone, that is constructional features, subsequently stripped of the weaker rocks which buried them, for example argillaceous limestone in east central Ireland, whereas some which underlie an unconformity are probably subaerially eroded features that have been similarly uncovered and represent true exhumed karst, for example the knolls of the western Peak District such as Chrome Hill.

Without having been buried, some karst forms are *relict* from previous morphogenic conditions no longer operative; thus Gavrilović (1969) maintains that in southern Yugoslavia there is karst inherited from Tertiary tropical climate. The names palaeokarst and fossil karst have been applied in such cases also. The former agrees with usage of the wider term, palaeoform, but the latter is less satisfactory in that 'fossil' usually implies interment.

Another fundamental distinction lies between karst that is completely surrounded by impervious rocks and that which can drain directly to the sea without the intervention of different hydrogeological systems. The French call the former *karst barré,* which can be rendered *impounded karst* as opposed to *free karst*. The impounded condition has most consequences when the karst area is small and very much affected by runoff from surrounding non-karst terrain.

There are also landforms and even landform assemblages which resemble karst forms but are the product of different processes. This *pseudokarst* is found in sediments such as the finer pyroclastics and loess as a result of piping. Here subsurface mechanical eluviation of clay particles leads to cavitation and to subsidence. Small caves may result such as those on Officer's Cave Ridge, Oregon, in altered tuff.

More obviously pseudokarstic are larger caves of several types and collapse depressions which occur in volcanic regions through liquid segregation in layered lava and so also are the caves, natural bridges, and melt dolines in glaciers formed by melting of ice *(glacier thaw karst)*. *Thermokarst*, in which melting of ground ice (permafrost) in unconsolidated sediments occasions subsidence and other forms, is less readily distinguished from true covered karst. These different kinds of pseudokarst are discussed in Ollier (1969) and in Davies (1969) in this series.

This label of pseudokarst has been applied to rounded grooves in granite, reaching largest dimensions in the humid tropics. The exact nature of the processes fashioning these *Silikatkarren* is not known but despite the low solubility of granite it is probable that they are due to solution. Even chemically inert rocks can develop minor features convergent with those of karst as exemplified in quartzite in the Wonderland of the Grampian Ranges, Victoria, and in the Carrao basin, Venezuela (White and others 1966). Priesnitz (1969) makes a strong case for the inclusion of all these in karst proper.

However far back in time one traces embryonic geomorphological notions, karst participates throughout the historical development of the subject. A bronze engraving records graphically the results of the first known expedition to investigate karst phenomena sent

by the Assyrian King Salmanassar III to the springs and caves at
the source of the Tigris River; the writings of several Greek and
Roman classical poets, philosophers, and natural historians des-
cribe karst forms from the limestone which abounds in the
Mediterranean. As in most matters the Middle Ages added little
to the knowledge of the ancients although inscriptions in the
famous Postojna Cave in Slovenia date from the thirteenth
century onwards. With the Renaissance direct observation of
nature came to supplement quotations from the classics, though it
was the economic search for minerals in caves which led to most
of the new knowledge. The seventeenth century, the century of
the first scientific societies, saw also the appearance of books
devoted to karst, the first being that of Jacques Gaffarel of Paris
in 1654, of which little survives unfortunately. In 1689 the
impressive work of J.W.F. von Valvasor on the Slovenian karst
appeared with many maps of caves. The next developments in
karst study lay in the less central direction of investigating the
fossil content of cave deposits, culminating in the writing of
G. L. Cuvier and the *Reliquiae diluvianae* of R. Buckland. The
first really systematic geological and hydrological investigation of
a karst area may be attributed to Adolf Schmidl on the Postojna
region about the middle of the nineteenth century.

The classical period in the development of geomorphology of
the second half of the nineteenth century and the beginning of the
twentieth was equally fruitful in karst investigation and in the
emergence of scientific societies and caving clubs to prosecute it.
In this time Édouard A. Martel's immense field and literary efforts
brought France alongside Austria in the forefront of karst studies
where it has since remained, and he also stimulated the English
into serious activity. Franz Kraus's *Höhlenkunde* (1894) marked
an epoch in the German literature and in 1893 appeared *Das
Karstphänomen* of Jovan Cvijić, the greatest of the Yugoslavian
karst geomorphologists. A. Grund's *Die Karsthydrographie*
(1903) provoked the most bitter phase of controversy over
underground water circulation in karst, which had older roots and
which persists in milder form to this day. In Czechoslovakia,
K. Absolon is chiefly remembered for the stimulus he gave to
Moravian karst investigation, whereas J. V. Daneš from Bohemia
was the first karst specialist to range the world in this pursuit, and
pioneered the study of tropical karst. In Belgium, E. van der
Broek, in Romania, E. G. Racovitza, and in Italy, O. Marinelli,

were in this period the most significant contributors in their respective countries to this flowering of the subject.

After World War I karst studies greatly intensified in countries where they were long established and spread to many others so that it becomes even more invidious to mention individuals. Institutes solely for karst and cave research came into being now —at Vienna, with which the name of Georg Kyrle, the author of *Theoretische Speläologie* (1923) and perhaps the first university professor of speleology in the world, is closely linked, at Cluj in Romania founded by Racovitza, and at Postojna, at this time part of Italy, where G. A. Perko was director. Only since World War II has there been a real spread of scientific study of karst away from the main centres of Western learning. International congresses of speleology began in 1953 and their proceedings are publications of rapidly growing importance. Between 1954 and 1964 the International Geographical Union had a standing Commission on Karst Phenomena under the chairmanship of Herbert Lehmann of Frankfurt am Main, who had fostered the climatic morphology approach to karst since 1936 with his study of Javan limestone relief. The Frenchman, Jean Corbel, was also an innovator in this aspect, particularly in the measurement of rates of limestone removal in varying climates. During and since World War II there has been great elaboration of the techniques of cave exploration, one of the necessary bases for the understanding of karst. The major stimulus here came from the French with the name of R. de Joly outstanding.

KARST LITERATURE

This brief history will have made evident one of the several difficulties about karst literature for the English speaker, namely, that the contribution of the Anglo-Americans has not loomed as large in this field as in practically every other aspect of geomorphology. In recent decades a surge of interest in the scientific side of karst has enlivened the older sporting interest in Britain and the United States, and has begun to redress this balance. However, it is still true that very many of the important books and serials are not in English. Moreover the really significant literature ranges beyond French and German into languages less well known to English speakers to a degree not common in science at large.

In particular, by reason of the richness and of the long history of investigation of the Dinaric karst, an important sector of writing in the south Slav languages lies behind a very real language barrier. Fortunately J. Cvijić's mature view of the subject is available to us in a posthumous work (1960) in French and a most helpful critical review of the Yugoslav contribution for most of the time since Cvijić's death has been prepared by A. Blanc (1958). A third difficulty arises from the fact that a great proportion of important results has appeared and is appearing, outside the geological and geographical serials familiar to the geomorphologist, in speleological literature, less readily available in libraries, often local in approach and one in which many journals have suffered great variations in effectiveness and too frequent early demise. Books on speleology are usually easier of access and good introductions to karst are found within their pages, e.g. Cullingford 1962; Trombe 1952; Moore and Nicholas 1964; Gèze 1965; Trimmel 1968.

TERMINOLOGY

Additionally it must be admitted that no branch of geomorphology is so troubled by terminology as this. No other kind of terrain seems to have bred such a multiplicity of regional terms so that reaching agreement on a common vocabulary within the frontiers of one country or one language is a major task in itself prior to that of securing international equivalence. However, vocabularies thrashed out by representative committees have been published for German terms (H. Trimmel, ed., *Speläologisches Fachwörterbuch*, 1965) and for French (P. Fénelon, ed., *Vocabulaire français des phénomènes karstiques*, 1968). In English the situation is less formalised but there is a useful glossary in *British Caving* (Cullingford 1962) and in a more convenient form we now have *A Glossary of Karst Terminology* by W. H. Monroe (1970). FAO is preparing a glossary of karst hydrogeological terms.

METHODS OF INVESTIGATION

Most field and laboratory methods in geomorphology find application in the realm of karst. Additional techniques are commonly employed in the study of caves and karst hydrology.

The survey and cartography of caves requires some adaptation of surface methods (Butcher and Railton 1966; Symposium on Cave Surveying 1970). Radiolocation has come into use as a check on cave surveying from the surface, and geophysical methods such as soil resistivity and gravimetry may help in the discovery of new caves. Cave diving is also a specialised technique, more valuable for the way it permits examination of completely waterfilled caverns where morphological development goes on in different fashion than for the access it gives to ordinary caves beyond water barriers.

The tracing of underground drainage connections is special to karst study and a variety of means of tagging water has been employed—salts, dyestuffs such as fluorescein, rhodamine and pyridine, and exotic spores (Drew and Smith 1969; Symposium on Cave Hydrology and Water Tracing 1968). The use of radioactive tracers presents health hazards but naturally occurring 'environmental isotopes' are now being put to use. Determination of other natural chemical and physical parameters of karst waters has long been used comparatively to ascertain underground links, but is now more valuable in establishing the rate and distribution of solution (Douglas 1969). Artificial flood pulses as well as natural flood pulses are now being monitored as a tool in making inferences about unknown systems behind karst springs.

II

KARST ROCKS

Many unsolved problems in karst geomorphology may arise from inadequate appreciation of differences between the rocks involved (Sweeting 1968). There is more variety under 'limestone' than any other common rock name. The usual definition is that at least half of the rock is made of carbonate minerals, mainly in the form of calcite ($CaCO_3$). This is satisfactory for karst study since the 60 per cent of $CaCO_3$ estimated by Corbel (1957) to be necessary for any significant karst development is probably of the right order. He also considered that a purity of 90 per cent or better is necessary for its full development, though this does not always result as with chalk with its 95 per cent content of $CaCO_3$. These points indicate the need for more detailed knowledge of lithology than is usually accorded karst studies.

Besides calcite, the minerals common in carbonate rocks are aragonite, dolomite, magnesium carbonate, and the impurities— chalcedonic silica (chert and flint), detrital and authigenic quartz, authigenic feldspar, and clay minerals (chiefly illite and kaolinite). Lesser impurities include siderite and derived oxides of iron, glauconite, collophane, pyrite, and bituminous matter. Quartz and clay are the commonest impurities. The limits adopted to define mixed rocks vary; those of Leighton and Pendexter (1962) are set out in Fig. 1. Marl is an ambiguous term used both for calcareous clays and very pure lacustrine limestones. Magnesian limestone is generally taken to mean limestone with an unusually high content of magnesium carbonate (though still never more than a few per cent).

LIMESTONE

Although terrestrial and freshwater limestones are locally

10

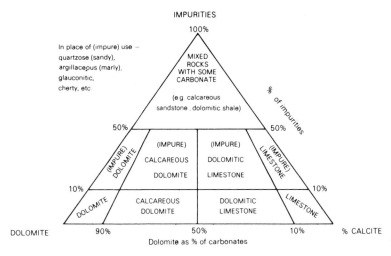

1 Limits for mixed carbonate rocks of Leighton and Pendexter 1962

important and in Australia loom large even on the continental scale in the form of aeolian calcarenite (calcareous dune limestone), most limestone is of marine origin. Many materials—detrital, chemical, and organic—go to make them up. They frequently undergo much change after deposition at low temperature and pressure (diagenesis) and may also be altered by high temperature and pressure (metamorphism) to marble, which is made up of large, clear calcite grains. Because of this complexity, there have been many attempts to classify them. Modern classifications hinge mainly on texture and as this has considerable bearing on response to karst processes, the following brief account rests on those of Folk (1959) and Dunham (1962). The ideal procedure would be to devise a special classification solely for geomorphic purposes.

A few limestones are the result of organisms growing together so that their calcareous skeletons are bound into a rigid framework. The voids between may be filled later on by materials washed in or by chemical precipitates. These autochthonous fossiliferous limestones are called *biolithites* by Folk (his Type IV, Fig. 2) and *boundstones* by Dunham. According to their dominant life form, they may be named algal, coral, stromatoporal, or bryozoal. In addition they assume two major dispositions—compact massive

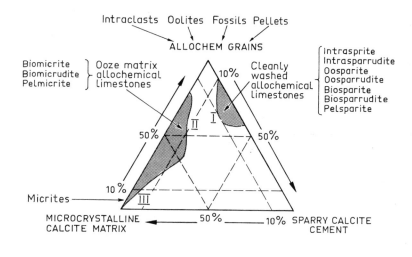

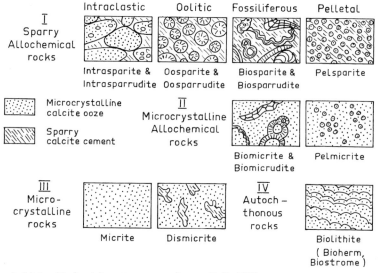

2 *Major kinds of limestone according to Folk 1959*

mounds or thick lenses which are called *bioherms*, and extensive tabular deposits or *biostromes*.

At an opposite extreme, there are limestones formed entirely or nearly so of lime mud or ooze, accumulating more or less in place on calm, shallow sea floors. These are called *micrites* by Folk (Type III, Fig. 2) and *lime mudstones* by Dunham. These microcrystalline rocks are the aphanitic or lithographic limestones, and calcilutites of older classifications. The lime mud may be disturbed whilst still soft by slumping, and by boring animals; voids so created get filled with precipitated sparry calcite. These rocks are Folk's dismicrites.

Very many limestones are more complex than these end members and consist of a framework of larger grains with a matrix in the intergranular voids, either of dull lime mud or of sparry (i.e. clear and lustrous) calcite cement.

The larger grains are transported or secreted aggregates, Folk's *allochems,* of which he recognises four important classes—intraclasts, oolites, fossils, and pellets. *Intraclasts* are detrital fragments, eroded from weakly consolidated carbonate sediments and included after transport in almost contemporaneously forming limestone. These fragments are usually rounded and range from sand to boulders in size. *Oolites* are spheroidal precipitates of concentric or radial structure around a foreign nucleus, forming where bottom currents are strong. *Fossils* in these rocks are not bonded in growth but consist of discrete skeletal components such as foraminifera, sponge spicules, corals, bryozoans, brachiopods, and molluscs. Algae such as *Lithothamnion* are the most important of all. Many of these organisms introduce aragonite into the rocks but it soon inverts to calcite over geological time. *Pellets,* well sorted and rounded aggregates of microcrystalline calcite, are thought to be faeces of worms and other invertebrates.

Where bottom currents were weak or impersistent, lime mud accumulated along with the larger framework constituents. In these rocks (Folk's Type II, Fig. 2) there is a great range in the proportion of allochems to muddy matrix. In some the grains are so far apart that they float in the mud, they are mud-supported; these are the *lime wackestones* of Dunham's classification. When the grains are abundant enough to be in contact and support one another, the rock is the *lime packstone* of Dunham.

Elsewhere currents were strong enough to pile the intraclasts, oolites, pellets or fossils together and to winnow out the ooze. These are all grain-supported rocks but the voids between may be partly or fully filled in later by the precipitation of clear calcite.

These rocks comprise Folk's Type I (Fig. 2) and Dunham's *lime grainstones*.[1]

Both the cleanly washed and the muddy allochemical limestone groups can be subdivided on the basis of their dominant framework constituents—intraclastic (detrital), oolitic, biogenic (skeletal), and pelletal. Also they can be categorised in terms of the dominant size of their large grains into the sand-sized calcarenites and the gravel- and boulder-sized calcirudites.

To cover all possible combinations concisely, Folk has introduced a set of terms compounded of two or three elements. The first element (intra-, oo-, bio-, pel-) depends on the proportions of the framework constituents; the second element refers to the nature of the material filling the voids in between (-micr-, if it is lime mud; -spar-, if it is sparry calcite); the third element (-rud-) is only used when the framework grains are larger than 2 mm, otherwise a calcarenitic texture is implied. All the terms end in -ite. Thus a clastic limestone consisting of oolites cemented by sparry calcite is an *oosparite*, whilst a limestone composed of bivalve shells of the order of 1 cm set in a matrix of lime mud is a *biomicrudite*. The dominant fossil type may be added where appropriate, e.g. crinoidal biosparite.

Folk does not include in that part of his classification limestones formed of clasts of much older, consolidated calcareous rocks; he designates these *calclithites* but thinks they are very rare.

Recrystallisation of both mud matrix and framework constituents complicates the classification set out and indeed may have proceeded so far that original depositional textures are no longer recognisable, except for some 'ghosts' of the allochems. Dunham terms these simply crystalline limestones. Marble is often loosely used for such unmetamorphosed rocks.

REEF FACIES

Many marine limestones have been laid down in association with 'coral' reefs—wave-resistant projections from the sea floor built

---

[1] The definitions of Folk's sparry allochemical limestones and Dunham's lime grainstones do not coincide exactly. Neither authority isolates limestones consisting entirely of allochems and voids. Yet these are highly porous and likely to behave in a special way in karst. Coquinas—shelly limestones—fall into this category.

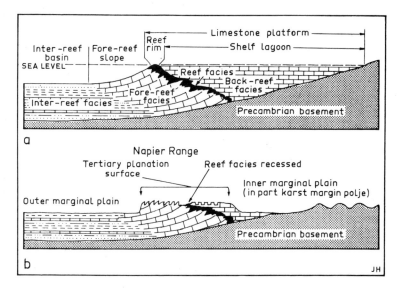

3 (a) Devonian reef facies in Napier Range, West Kimberley, Australia. After Playford and Lowry 1966.
(b) Present relief.

by colonial organisms, and characteristic facies patterns of different lithologies develop in such contexts (Playford and Lowry 1966).

The reefs themselves, whether atolls in oceanic situations, linear and patch reefs in barrier complexes or fringing reefs, are massive bioherms with varying fossiliferous composition (Fig. 3). However, the bioherms often form only a low percentage of the whole reef structure. Cavities in them are numerous initially but these tend to fill with intraclasts, ooze, fossils, or terrigenous sands. Recrystallisation is common and so is replacement of the aragonite of many organic skeletons by calcite. Dolomitisation is also characteristic but irregularly patchy.

The steep seaward flanks are usually subject to violent wave action and here submarine talus accumulates in forereef beds with steep primary or depositional dips up to 30-35°. Characteristically this forereef facies consists of intrasparrites and intrasparrudites. The steep submarine slopes are unstable and subject to slumping when still only partly lithified. This process gives rise to the *megabreccias* of Playford and Lowry (1966)—intrasparrudites with many large blocks greater than 3 m across.

On the protected sides of the reefs, particularly in lagoons, the backreef facies accumulates as nearly horizontal, well-bedded deposits. Close to the reef, clastic material from it may be washed over to accumulate at modest dips as intramicrites. Biostromes are especially characteristic of the backreef facies and there are biomicrites, oomicrites, and micrites through the prevalent accumulation of calcareous oozes in the quiet conditions.

These different facies in reef formations can be important in karst areas as is evident from the Permian reef complex of the Guadalupe Range in New Mexico, U.S.A., which includes Carlsbad Cavern, and in the Devonian reef complexes of the Limestone Ranges of West Kimberley, Australia.

DOLOMITE

Most dolomites appear to be secondary in origin through the replacement of limestone; consequently they are dominantly crystalline. They are best classified on the basis of the mode of crystal size and of the nature of the surviving determinable constituents from the limestone, which are usually the 'ghosts' of intraclasts, fossils, oolites, or pellets. It is impossible to determine the original proportions of ooze and sparry calcite. Dolomites with crystals finer than 10 μ, i.e. dolomicrites, lacking relict limestone textures and associated with evaporites, are thought to be of primary origin, probably forming in very warm, hypersaline lagoons or embayments.

EVAPORITES

Similar in origin to primary dolomite are the evaporites. There are many evaporite minerals and these occur in complex mixtures as rocks. However, easily the most important are halite (rock salt), anhydrite, and gypsum.

Halite (NaCl) is a massive, coarsely crystalline rock, lacking joints. It yields plastically at low pressure and temperature; consequently it frequently occurs in the form of diapiric 'salt domes'. Clay and anhydrite impurities are usual.

Anhydrite ($CaSO_4$) is more frequently finely granular than crystalline and occurs in thick beds or may be thinly laminated.

Gypsum ($CaSO_4.2H_2O$) is formed by hydration of anhydrite, during which it swells 30-50 per cent and much crumpling may result. It is usually finely granular, may be thin bedded or very massive and jointless.

In the Permian Castile Formation of New Mexico and Texas. all three occur together, accumulating to an exceptional thickness of over 1200 m.

TABLE 1 **Permeability of common rocks**

Rock	Range of permeability coefficients ($l/sec/m^2$ with hydraulic gradient of 1 in 1)		
Unconsolidated:			
Clay	0·35	—	35
Sand	350,000	—	350,000,000
Gravel	350,000,000	—	350,000,000,000
Indurated:			
Shale	0·0035	—	3·5
Sandstone	35	—	3,500,000
Conglomerate	35	—	3,500,000
Limestone	**3·5**	**—**	**350,000**
Basalt	3·5	—	35,000
Granite	0·0035	—	0·35
Gneiss	0·0035	—	3·5
Tuff	0·35	—	35,000

Based on J. P. Walz, Ground water. Pp. 259-67 in *Water, Earth and Man,* ed. R. J. Chorley. London, 1969.

PORES AND PLANES OF WEAKNESS: PERMEABILITY AND
STRENGTH

The response of karst rocks to geomorphic processes depends very much on their permeability and mechanical strength as well as their purity, and these in turn hinge in large measure on porosity and planes of weakness. These interrelated factors will be discussed together for convenience and brevity.

Porosity refers to all the voids in a rock expressed as a fraction or percentage of the bulk volume and is determined on specimens in the laboratory. Permeability on the other hand is the capacity

TABLE 2 **Compressive strength of common rocks**

Karst rocks	Uniaxial compressive strength (bars)	Other rocks	Uniaxial compressive strength (bars)
Limestone (excluding chalk and coral rock)	340 – 3310	Shale	360 – 2310
		Sandstone	120 – 2350
		Conglomerate[a]	1660
Limestone, reef breccia	60 – 350	Tuff	350 – 2620
		Basalt	810 – 3590
Marble	460 – 2380	Diorite	1550 – 3350
Dolomite	620 – 3600	Granite	1590 – 2940
Anhydrite[a]	410	Gneiss	1530 – 2510
		Quartzite	1460 – 6290

Suggested limits for terms

Very weak	< 350
Weak	350 – 700
Strong	700 – 1750
Very strong	> 1750

[a] Single rock sample

Based on D. F. Coates, *Rock Mechanics Principles.* Mines Branch Monograph 874, Ottawa, 1967; J. C. Jaeger and N. G. W. Cook, *Fundamentals of Rock Mechanics.* London, 1969; L. Obert and W. I. Duvall, *Rock Mechanics and the Design of Structure in Rock.* New York, 1967.

of the rock to transmit water and it is determined in the field from bores and wells as well as in the laboratory. Table 1 sets out the coefficients of permeability for some common rocks. These measures vary in relation to the size and continuity of the voids, and limestones range most drastically in both the controlling factors and permeability.

The mechanical strength of rocks can be determined in various ways in the laboratory. Table 2 sets out the uniaxial compressive strength of some common rocks; this is the maximum stress cylinders of rock can withstand under compression in a single direction without rupture or bending. Tensile strength, the maximum tensional force rocks can undergo without deformation, is much harder to measure and varies much more from one specimen to another. However, uniaxial compressive strength is usually a reasonable guide to variation in other strength properties of rocks, including the tensile strength, which is, of course, many times less than it. It will be seen from Table 2 that karst rocks, especially limestones, are exceedingly variable in mechanical strength, varying from very weak to very strong. But the behaviour of large masses of rock in nature is even more variable than these laboratory determinations suggest because of larger scale inhomogeneities, especially the planes of weakness traversing them.

Intergranular porosity (sometimes called 'primary permeability') is a textural characteristic of the rock closely related to its sedimentary origins and diagenesis, and any later metamorphism. Biolithites, skeletal and detrital limestones of many kinds can be extremely porous because of the irregular shapes of many fossils and some intraclasts, whereas micrites are usually devoid of pores. However, initial pores may be filled completely with lime mud or calcite cement, and very many limestones have virtually no porosity of this type. Dunham (1962) points out that grain-supported limestones are more likely than mud-supported limestones to acquire intergranular porosity after deposition as a result of differential solution and compaction in several ways. If pore infilling does not take place during deposition or diagenesis, it may well occur subsequently through recrystallisation (tending to crystalline carbonates) so that the older the limestone the more likely it is to be non-porous. Martel (1921) gives an intergranular porosity range of 0·09 to 2·55 per cent for the 'limestones' of France and 14·4 to 43·9 per cent for the chalks (in northern England chalk can reach 46 per cent, Kendall and Wroot 1924).

The mechanical strength of limestones is closely related to their porosity and consolidation; it greatly affects surface slope development in karst and even more the possibility of significant cave development. Weak limestones may collapse into incipient caves and this may in turn affect the nature of the hydrologic system. This factor probably dominates the evolution of chalk country. Chalk is mainly a pure biomicrite, very porous and of earthy nature, lacking strength. It develops into gently rounded relief, with very rare crags at the points of pronounced river or marine attack, characterised by extensive dry valley systems and few caves, though some are known in both France and England. Water infiltrates rapidly into chalk surfaces but does not escape so readily; underground circulation is dependent on joints as in most limestones (Brown 1969; Ineson 1962).

The fewer the intergranular voids, the more cemented and

compacted the limestone, the more important become its planes of weakness, which promote permeability through their solutional enlargement ('secondary permeability' of some). In fact the nature, frequency, and pattern of planes of weakness probably constitute the most important single factor of structure (in the geomorphic sense) in karst. Indeed Tricart (1968) has argued that a combination of high intergranular porosity and poverty in joints is responsible for poor karst development in Barbados despite very soluble rock, heavy rainfall, and lush vegetation, through failure to canalise underground circulation.

Bedding in limestone may involve (1) dividing surfaces only, (2) variation from top to bottom of beds, with greater purity at the bottom, or (3) lithologic differences between succeeding strata. Cyclic changes in the beds are common in limestone sequences. Thin bedding is usually unfavourable to karst because it is often accompanied by insoluble terrigenous constituents in the parting planes and by shale or clay interbeds, which may block incipient cave and underground drainage development. Moreover, it may also lead to reduced strength with equivalent effect. With thick beds, the bedding planes are conducive to cave development. In the English Peak District, according to Warwick (1953), caves are preferentially fashioned in massive bioherms.

Joints are planes of weakness favourable to karst, though rewelding of joints by solute precipitation to minimise their geomorphic effect is common in carbonate rocks. In these rocks, joints are usually due to release of strain energy, residual from earlier compression, during later uplift which permits extension under load (Price 1959; Hancock 1968). Because of this dominant origin, most joints are normal to bedding planes. Minor joints lie within a single bed whilst major joints cut several. They usually occur in parallel sets and frequently two sets intersect, commonly at about 60°, in a conjugate joint system (Pl. 9). Joints are (1) latent—thin and hairlike, only apparent when the rock is broken up, (2) closed—visible and allowing descent of capillary water, and (3) open—allowing free water movement.

Joints may be too close set for adequate rock strength for cave formation. This may be particularly true of cleavage joints, close-set planes of weakness at right angles to compressional forces, which usually make the rocks too weak mechanically for good cave development as at Shatter Cave in the Goodradigbee River valley, New South Wales, so named because of its instability.

1 *Stylolites weathered out to give appearance of flaggy bedding. Lower surface smoothed by subsoil solution and exposed by soil removal. Oligocene limestone, King Country, New Zealand.*

Opposite views have been expressed about the role of faults in karst development (Stringfield and LeGrand 1969a). They are frequently accompanied by much cleavage; mineralisation is another accompaniment which will influence their geomorphic role. In soft chalk, plastic material injected into a fault zone can make it a partial barrier to groundwater flow (Ineson 1962). Factors such as these may be responsible for their varying effects on underground drainage, doline and cave development.

Stylolites may be so prevalent in limestones and dolomites as to affect their relief expression significantly. These are sutures in the rock where pressure solution has taken place, often leaving thin laminae of insolubles. The solution may have taken place along bedding planes or elsewhere in the rock. Recementation takes place often enough and the stylolites may become the resistant parts of the rock (Pluhar and Ford 1970). The widespread Oligocene shell calcarenites of New Zealand have many residual seams of this nature as a rule (Barrett 1963), and this gives rise to a ribbing in outcrop, which appears geomorphically to resemble lensy, flaggy bedding, with the bedding planes projecting (Pl. 1).

III

KARST PROCESSES

Karst as defined in this book—namely as a type of landscape with distinctive landforms that arise primarily from abnormally high solubility of the bedrock—includes terrains in which processes such as mechanical action of rivers and frost shattering have played significant, even dominant parts, though these processes are in no way special to karst. By employing such a definition, unrealistic standpoints are avoided such as that of Panoš and Štelcl (1968) who excluded areas such as the Sierra de los Organos, Cuba, from karst, even though solution played a particularly great role in producing distinctive landscape there. Nevertheless, only those processes or aspects of processes assuming peculiar importance in karst will be discussed here.

SOLUTION AND PRECIPITATION

Limestone

Since limestone is the most widespread karst rock, its solution and deposition are the most important processes to consider. As calcite, it is only modestly soluble in pure water, at saturation the amount varies from about 13 mg/l at 16°C to about 15 mg/l at 25°C (Frear and Johnston 1929). Aragonite, a much rarer mineral, is about 16 per cent more soluble.

However, much greater concentrations than these are common in natural waters. Other solutes are responsible for this and varied lines of evidence mustered by many authors have shown that usually the most important is carbonic acid (e.g. Adams and Swinnerton 1937; Smith and Mead 1962; Gross 1964; Pitty 1966).

Much less work has been done on the role of organic acids resulting from rotting of vegetation such as lactic acids. Yet there can be no doubt about the importance of organic acids regionally as Williams (1970) has pointed out in relation to the widespread contact of peaty water with limestone in Ireland, and perhaps generally (Muxart and others 1968). Chelates are other organic compounds that have an action similar to that of acids on limestone (Keller 1957).

Of more localised importance, sulphuric acid produced by the weathering of sulphide minerals such as pyrite and marcasite, sometimes from interbedded shales, is a powerful solvent of limestone (Morehouse 1968; Pohl and White 1965). Another source of sulphuric acid may reside in the physiology of certain iron-fixing bacteria (e.g. *Crenothrix, Callionella*). Nitric acid in rainwater as a result of lightning is apparently negligible as a limestone solvent.

Because of the paramount part played by carbonic acid, our main concern is with the carbon dioxide-water-calcium carbonate system. This will be set out here in a simplified way though it has a much more complicated physical chemistry than is evident from most geomorphology texts. It is not just a matter of a single reaction to produce more soluble bicarbonate from less soluble carbonate but of a series of reversible reactions and ionic dissociations each governed by different equilibria (Bögli 1960; Roques 1964, 1969; Thrailkill 1968).

Dissolved calcium carbonate is in an ionic state:

$$CaCO_3 \text{ (solid)} \rightleftharpoons Ca^{2+} \text{ (hydrated)} + CO_3^{2-} \text{ (hydrated)} \quad \cdots (1)$$

and the product of the two kinds of ion is a constant. The warmer the water and the more agitated its motion the less time solution takes to reach the saturation equilibrium values already given; the nature of the limestone surface will also govern the rate of solution. For greater amounts of limestone to be taken into solution than the equilibrium for the temperature, a further chemical reaction has to take place with carbonic acid.

Natural waters have some carbon dioxide in solution; the equilibrium amount increases with rising partial pressure of carbon dioxide (Pco_2) in the air in contact with the water and falls with rising water temperature. As temperature rises from $0°$ to $35°C$, the amount at saturation drops between one-third and two-thirds. These relationships constitute Henry's Law. A small part of the dissolved CO_2 reacts very rapidly with the water to produce carbonic acid which is always in an ionic state:

$$CO_2 \text{ (dissolved)} + H_2O \rightleftharpoons H^+ + HCO_3^{2-} \quad \ldots (2)$$

Carbonate ions from the dissolved limestone react instantaneously with the hydrogen ions to produce bicarbonate ions:

$$CO_3^{2-} + H^+ \rightleftharpoons HCO_3^{2-} \quad \ldots (3)$$

This last reaction upsets the equilibria of (1) and (2). More limestone goes into solution to keep the product $[Ca^{2+}] [CO_3^{2-}]$ constant and more dissolved carbon dioxide reacts with water to produce more carbonic acid. This phase is rapid but it only achieves a modest additional solution of about 8 mg/l in normal atmosphere.

However, this last phase produces a disequilibrium between the carbon dioxide partial pressures of the air and of the water. There follows a diffusion of CO_2 from the air to the water, in turn permitting further solution of limestone through the chain of reactions. The complete process of limestone corrosion may therefore be simplified in the following way, reminding us of the various stages:

$$CaCO_3 \text{ (solid)} + H_2O + CO_2 \text{ (dissolved)} \rightleftharpoons Ca^{2+} + 2HCO_3^-$$
$$\Updownarrow$$
$$CO_2 \text{ (air)}$$

Diffusion of CO_2 through water is a slow process compared with the earlier stages. It varies with temperature, proceeding faster the higher the temperature, but more important is the speed and turbulence of water movement (Weyl 1958). These are thus the chief controls of the rate at which limestone solution takes place. However, this final phase of CO_2 diffusion can only happen where the water is in contact with air (an open system); it cannot do so where water is confined between rock alone (a closed system).

The total amount of limestone which can be dissolved at saturation equilibrium per unit volume of water is overall a direct function of the carbon dioxide partial pressure of the air with which the water is in contact and an inverse function of the water temperature because of the latter's control of the dissolved CO_2 saturation equilibrium (Henry's Law), see Fig. 4. The temperature effect is, however, relatively modest—about a threefold range results, whereas the partial pressure effect is much greater—at least a hundredfold range (Ek 1969). In the free atmosphere, the

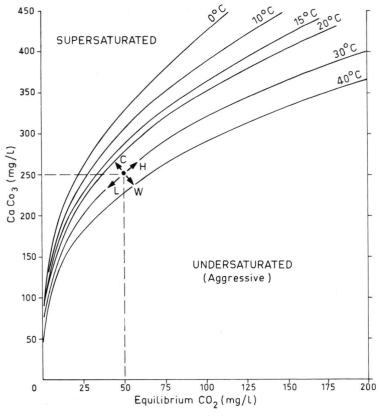

4 Saturation equilibrium curves for solution of calcium carbonate at different temperatures as a function of equilibrium carbon dioxide in solution. After Trombe 1952.

If saturated water at 250 mg/l CaCO₃ and 50 mg/l CO₂ is cooled (direction C), it can dissolve more limestone; if it is warmed (W), it will throw some out of solution. If it comes into contact with air with less CO₂ (L), it will lose some CO₂ and precipitate CaCO₃; if it encounters air with more CO₂ (H), it will take in more CO₂ and dissolve more limestone.

mean figure for P_{CO_2} is quite small, 3×10^{-4} bar (0.03 per cent of volume). Most cave air has similar values but it does vary significantly (Holland and others 1964; Ek and others 1968). The air in snow has about 0.1 per cent CO_2 because the smaller oxygen and nitrogen molecules diffuse out of the voids faster than the larger CO_2 molecules. Theoretically therefore snow and glacier ice meltwater could be richer in CO_2 than rainwater. But the really significant amounts of CO_2 are found in soil air, including that in vegetation litter. Values of 1-2 per cent are usual, but very much higher values occur. Extreme quantities of 20-25 per cent have been reported from poorly ventilated tropical soils. Root respiration and bacterial decay of organic matter seem to be chiefly responsible for this.

'Biogenic' carbon dioxide is therefore regarded by most investigators as the prime control of ultimate limestone solution per unit volume of water and it in turn is chiefly dependent on temperature and rainfall which promote vegetation growth. Smith and Mead (1962) have shown that the springs of the Mendip Hills in southwest England have taken an average of 280 mg/l of calcium carbonate into solution and this agrees with saturation equilibrium at prevailing temperatures for the average figure of 1·6 per cent carbon dioxide in the soil air beneath the grass pasture of these hills.

The carbon dioxide-water-calcium carbonate system so central to an understanding of karst has some important corollaries. If water has reached saturation with respect to limestone, either at the surface or at shallow depth underground, descends further and is cooled in the process, as may happen in summertime in particular, it becomes capable of dissolving more carbon dioxide

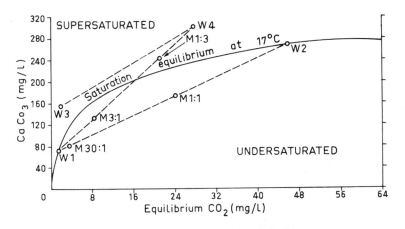

5 *Mixing corrosion. After Bögli 1964a and Thrailkill 1968.*

Mixing of saturated waters W1 and W2 results in aggressive water. M30:1 (30 parts low Pco₂ W1 and 1 part high Pco₂ W2) is more undersaturated than M1:1, an equal mixture. No mixture of supersaturated waters W3 and W4 will be aggressive. M1:3 (1 part saturated W1 with 3 parts supersaturated W4) remains supersaturated but M3:1, a converse mixture, becomes aggressive.

from cave air and thus of becoming aggressive to limestone once more. Bögli (1964a) has termed this cooling corrosion.

However, he regards as much more important another mechanism which he terms mixing corrosion (Fig. 5). Because the relationship between CO_2 partial pressure and calcium carbonate saturation equilibrium is an exponential one, the mixing of two saturated bodies of water with different calcium bicarbonate concentrations produces water which is undersaturated. It then becomes capable of renewed attack on limestone with which it is in contact. This mixing is most effective in reviving aggressiveness when a large body of water saturated in response to low P_{CO_2} mixes with a small body of water saturated at high P_{CO_2}. This may well be the case when vadose seepage and vadose streamflow (see Chapter V) mix. However, though the physical reality of mixing corrosion is not in doubt, it may not be as important geomorphically as Bögli thinks. For instance, one or both of the waters mixing may be supersaturated and little or no resumption of aggressiveness may result (Thrailkill 1968).

It is also misleading to ascribe too wide an importance to the saturation equilibrium of the overall solution process since very often natural waters will not equilibrate before escaping from many geomorphic situations. The amount of limestone dissolved is the product of the volume of water passing in a given time through any karst situation and its actual carbonate concentration, not the theoretical amount which could possibly be dissolved in those conditions. The water volume is ultimately governed by the surplus of precipitation over evapotranspiration and the $CaCO_3$ concentration usually depends on

1. the nature of the limestone surface;
2. the velocity and turbulence of the water flow;
3. the direct temperature effect on the rate of chemical reactions (though this factor is only significant when comparing karsts in climates very different thermally).

Less frequently will the P_{CO_2} and temperature controls of saturation equilibria be the arbiters from the geomorphic point of view.

Saturation comes into its own in the process of calcium carbonate precipitation. Many textbooks have stressed evaporation in this connection, which undoubtedly does raise concentrations above saturation equilibria to bring about deposition. But relative humidity in most caves is so high and air movement so slow that evaporation is there at a minimum. However, the various reactions of the CO_2-H_2O-$CaCO_3$ system are all reversible without the intervention of evaporation. If water has acquired high carbonate content in response to high P_{CO_2}, for example as a result of passing through soil and thence through tight rock fissures without air, it will diffuse the gas back into normal air if it resumes contact with the latter. This brings about carbonate super-saturation and precipitation of calcite. There can be little doubt that this is the dominant process in many contexts, especially in caves. Holland and others (1964) have shown this to be the case in Luray Caverns, Virginia, where saturated seepage water depositing calcite declines in calcium content whilst doing this but maintains its magnesium concentration. If evaporation caused the deposition, both magnesium and calcium would decline. A further but less important cause of deposition is the warming of water which has achieved saturation at lower temperatures; saturation equilibrium for CO_2 solution is lowered by this, diffusion of CO_2 occurs and calcium carbonate is precipitated.

Reprecipitation of $CaCO_3$ is important not only in giving rise to independent deposits and thus constructional landforms but can be equally significant in case-hardening limestones and calcareous sands (Monroe 1966; Jennings 1968). Induration by reprecipitation of calcite in voids near the surface gives calcrete or kankar which can be a vital factor in landscape development.

Dolomite

Dolomite rock behaves in natural waters in an essentially similar way to that of limestone, though the various equilibria in the solution of the double carbonate mineral, dolomite, which is its chief constituent, have been less fully investigated (Holland and others 1964). Under normal air and water interface conditions, it is claimed that dolomite is usually the less soluble, as is witnessed by recessive weathering of calcite veins in dolomite. But the situation is more complicated than this (Douglas 1965). With very high P_{CO_2} more magnesium than calcium is dissolved from dolomite as is also the case at very low P_{CO_2}, whereas at the pressures frequent in soil voids and rock crevices, they are dissolved about equally. The responses of subaerial and subsoil surfaces of dolomite should therefore vary. Magnesium carbonate which often occurs in carbonate rocks behaves in a complex way also because it occurs in three mineral forms, each with its own characteristic behaviour.

These complexities introduced by the presence of magnesium make study of the geomorphic role of solution in karst difficult because limestones are often partly dolomitic or magnesian. This can be illustrated by the attempts which have been made to infer from the Ca/Mg ratio of spring waters the nature of the rocks feeding the spring (Hem 1959; Jennings and Sweeting 1963). As we have seen this ratio will vary with the P_{CO_2} of the dissolving waters as well as with the lithology. Moreover the ratio also varies as a result of differential deposition, calcite preceding magnesium minerals in precipitation, whether this be the result of evaporation or of carbon dioxide diffusion to the air (Douglas 1965). Impurities such as salt (NaCl) and metal trace elements affect the carbon dioxide-water-carbonate system significantly in natural waters and this may explain many difficulties found in interpreting field measurements of carbonate content, temperature, and pH (Roques 1969).

Evaporites

The third most important karst rock, gypsum (Pl. 2), is much more soluble than either limestone or dolomite (Trombe 1952); CO_2 is not involved since it cannot react with either ion of $CaSO_4$. Gypsum is most soluble at 37°C, a temperature above that of most natural waters. Therefore what matters geomorphically is gypsum's increasing solubility with rising temperature up to that maximum. Conversely deposition results from the cooling of natural waters and by evaporation causing supersaturation.

Essentially the same relationships apply with the chloride rocks such as halite but these are more soluble still so that they can seldom persist in the surface landscape.

PIPING

In Chapter I reference was made to piping as a cause of pseudo-karst; it needs mention here because it also occurs in karst proper. Piping or tunnelling occurs in clastic sediments and soils where percolating water transports clay and silt fractions internally, leaving underground conduits (Parker 1964). It is thus related

2 *Steep gypsum karst below Pointe de la Jaquette, Bas Queyras, French Alps. Soluble rock yields active gullies and crumbling spires.*

more to eluviation than to solution, although the solution of soluble grains in a soil or sediment may assist piping.

It can occur in material of any grain size. In coarser materials the open fabric provides the necessary permeability. In fine-grained materials, extensive cracking and/or fine dispersion are necessary. It follows that the presence of clay minerals with high swelling capacities (e.g. montmorillonite), and a high dominance of Na and Mg over Ca in the exchange complex, are both conducive to piping in fine-grained materials.

Therefore piping occurs in various soils and superficial deposits on karst rocks. Indeed piping may be specially promoted by solution beneath the covers providing openings for removal of fines as well as by solution of residual rock fragments in some of them. Piping is accompanied by subsidence[1] with the development of shafts and surface depressions. Since solution and collapse[1] induce subsidence in covers, it is difficult to assess just how important piping is in karst but it must not be overlooked.

SUBSIDENCE

Karst is favourable in some ways and unfavourable in others to the mass movement of residual and transported mantles, and of interstitial bodies of soil and sediment. Antagonistic is the tendency of these materials to be drier than on impervious rocks; other things being equal, downward percolation into the bedrock allows covers to dry out more quickly and pronouncedly than on other rocks where this will depend more on runoff and lateral percolation. For this reason the more lubricated kinds of mass movement—block (rotational) slump, debris slide, debris avalanche, and debris flow (terminology of Varnes 1958)—are proportionately less active in modifying karst landscapes than others in the same climatic conditions. Because of the low percentage of insolubles in karst rocks, residual soils tend to be shallower on the average than on many other rocks, minimising

---

[1] In this book a distinction will be made between collapse—sudden mass movement of the karst bedrock—and subsidence—mass movement, often gradual, of soils, weathering mantles, and superficial deposits. With many rocks, this would be a highly artificial distinction but with karst rocks the dominance of solution usually ensures a sharp division in profile between bedrock and regolith.

superficial mass movement. Furthermore the ready reversal of the solution process leads to frequent cementation, reducing mobility, e.g. in talus.

Working against these factors is the effectiveness and widespread action of solution in removal of support in all kinds of unconsolidated materials. Removal of lateral support is clearly evident at springheads and at streamsinks, for example, though the infrequency of surface streams reduces the opportunities for undercutting of channel banks. Removal of underlying support is, of course, the outstanding characteristic of karst in this context, occurring where solution or collapse of the bedrock has undermined the covers at central points and induced centripetal mass movement. This frequently takes place as creep, block slump, and debris slide; debris avalanche, debris flow, and mud flow are proportionately less important for the reasons given above. Dry soilfall and slow earthflow on the other hand can assume unusual importance. For example, dry soilfall appears to be the dominant process in the Maiden's Tresses Chamber of Easter Cave near Augusta, Western Australia, where sandy loam soil has been fed vertically from a choked solution pipe in virtually single grain state to build a conical pile below. Very widespread is the gradual descent of surface materials in slow flow or slide down joints and other solution-widened planes in the bedrock, usually with a modest degree of lubrication. Surface clays and loams gradually incorporate blocks of the bedrock on their way down, producing gash-breccia, a very important category of cave sediment. There may be little or no surface evidence of the whole of this process.

Desiccation of covers is not entirely a restraining factor in mass movement because with swelling clays it readily creates deep cracking. Rapid infiltration from heavy rainfall down such cracks is a common trigger of block slumping.

COLLAPSE

In no morphogenic system is collapse—essentially rockfall, block slide and rock slide (Varnes 1958)—as significant as in karst for several reasons. Bare rock slopes and cliffs, where these operate, are more frequent since soils are thin. Solution acts as freely laterally as downwards to a degree not true of other erosive processes so that undercutting by streams is more effective in karst than in less soluble terrain. Moreover solution produces caves where roofs as well as walls are subject to fall. Nevertheless geomorphologists have shown little interest in this aspect of karst dynamics until recently (Davies 1949, 1960; Montoriol-Pous 1951, 1954; Renault 1967). Renault has ventured notably into the difficult task of applying the laws of rock mechanics to the complexity of natural situations in karst. In engineering laboratories and even in tunnel or cutting construction, stresses are applied instantaneously in terms of geological time whereas the deepening or widening of a valley or the enlargement of a cave proceed infinitesimally slowly in comparison; findings may not be easily transferable from the one domain to the other.

Rocks are not uniform lithologically nor are they isotropic structurally; bedding, jointing, and cleavage intervene. Therefore shear strength varies complexly. Nor is gravity the only force involved when a slope is undercut or a cave passage created; there are permanent tectonic stresses in the rocks as well. Also decompression by erosion opens up joints, which permit rockfall on the surface and roof collapse underground. Changes in the surface form or cave shape through solution, corrasion, and removal of cave deposits gradually upset a previously established equilibrium of forces until breakdown occurs to produce a fresh temporary equilibrium. Draining of a waterfilled cave system will reduce support quickly and this has been considered by Davis (1930) and others as a potent factor in cave evolution.

Stresses in cliffs in the open and on cave walls and roofs tend to drive them into space; falling and gliding of masses of rock result. In the confined condition of a cave, there is the difference that the continuity from wall to wall results in a local compression. Tensional and shear components in this compression tend to produce dome or arch forms, though how perfectly this will be achieved depends on the consolidation, planes of weakness and attitude of the rock and the pressure of the overlying rock, which depends on depth below the surface. Below a critical depth, the mechanical response of the rock will be different, a plastic regime taking over from an elastic one. Here an elliptical form with a vertical major axis should develop, though if collapsed material on the floor is not removed, this will interfere with the system of forces. This approximates to the interpretation Trimmel (1950, 1951) placed on the shape of certain Austrian caves in limestones

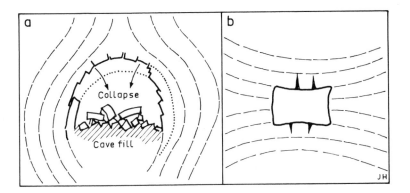

6 *Effects of stress on cave cross-section*
   (a) *Cave enlargement through spalling under high rock pressure at depth. After Trimmel 1968.*
   (b) *Stress zone around a shallow cave. After Davies 1960.*

of great mechanical strength (Fig. 6a). However, the critical depth is considerably more than 150 m postulated by Davies (1960) and plasticity will not be relevant in most caves (Renault 1967; Schoemaker 1948).

Halite and gypsum have greater plasticity than most carbonate rocks so that collapse will occur more readily in these rocks (Messines 1948). Chalk also can behave more plastically than most limestones. Whether this helps to explain comparative absence of caves in these rocks is not certain.

At shallow depths, arches are certainly produced by piecemeal collapse as in the chalky bryozoan limestone of the Wilson Bluff Limestone of the Nullarbor Plain, Australia (Pl. 3), and in weaker rocks still such as semi-consolidated dune limestone where domes form, with the space they enclose remaining almost full of collapse material (Fig. 7). With very strong rocks near the surface, the rock structure intervenes pronouncedly and collapse results in quadrangular cross-section with very low (McEachern Cave, Victoria: Ollier 1964) or very high dips (Pl. 38), and triangular forms with intermediate dips. When a cave is very near the surface, the roof behaves like a beam, which tends to sag into the opening and eventually collapse occurs, opening the cave to the sky (Figs. 6b, 38). Theoretically, prior to this release of forces, there should also be an upward compression of the floor of the cave. This has not been demonstrated in nature because

3 Roof dome and blockpile formed by breakdown in horizontally bedded Eocene limestone. Brackish water piped to surface for sheep. Koonalda Cave, Nullarbor Plain, South Australia. Photo by H. Fairlie-Cuninghame.

bedrock floors of any extent are rarely free from rockfall and other cave fill.

In terms of observable process, the patterns of tensional and shear stresses and of shear strength in cliffs and caves are for practical purposes permanent but the incidence of rockfall and rock slide is intermittent. Triggering actions are involved but little study has been made of them. Additional transitory stresses at the moment of earthquakes are obvious in this connection but there is probably no reason to think they play a role more important underground than on the surface or in karst than in other terrain. Rock blasting in quarries or for road construction or heavy vehicular traffic near caves could have this effect, though the Dark Cave, Bukit Batu, Kuala Lumpur, Malaya, shows no signs of collapse in the part exposed to regular tremor from a quarry crushing plant close by. Hydrostatic pressure of water in cracks may increase after rainstorms and act in this way. Other possible triggering forces are the swelling on wetting of colloids in slightly opened planes of weakness or of shale interbeds, and the force of crystallisation of minerals, especially of the commonest, calcite. Conversely it has been suggested that the drying out of clay interbeds or breccia fills in fissures may reduce

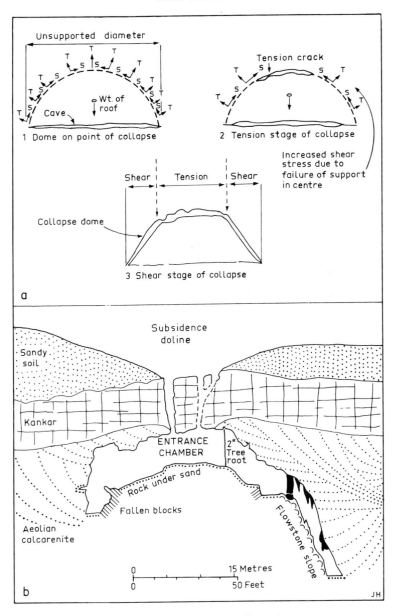

7 *Collapse dome formation in weak aeolian calcarenite*
    (a) *Sequence of collapse after A. L. Hill (in Jennings 1968). S = shear;*
        T = tension.
    (b) *Collapse dome in Easter Cave, Augusta, Western Australia. Soil*
        depths from Lowry 1967a.

cohesion and shear strength. With caves near valley walls, retreat of the latter decompresses the rocks and joints can open up as a result and promote cave collapse. However, the probability is that the commonest trigger will be the simple progress of solution by seepage water along planes of weakness and stream flow undercutting walls.

OTHER WEATHERING PROCESSES

There is no need here to consider surface weathering processes in karst other than the dominant and characteristic one of solution. Since many weathering processes are dependent on large and repeated changes in atmospheric conditions and on biological activity, the equable atmosphere and the comparatively abiotic conditions of most caves imply their absence from cave walls and roofs in large measure.

Departures from this generalisation may be important locally, however. Thaw-freeze action can penetrate into the forward parts of caves, acting direcdy on their surfaces, and its products can penetrate further still under gravity. Some caves are dry enough for crystallisation by evaporation and there salt weathering can play a major role in wall and ceiling sculpture, e.g. in some caves of the Nullarbor Plain (Lowry 1964; Jennings 1967c and its Fig. 2:12). In biological weathering, there are doubtful claims about bats scratching sizeable bell holes in cave roofs (King-Webster and Kenny 1958; contrast Hooper 1958) but there is the undoubted fact of corrosion of both bedrock and secondary cave deposits by bat guano and possibly urine (Jennings 1963). Moreover recent recognition of the ubiquity of bacteria has led to the idea that they may have both corrosional and depositional actions in caves (Smyk and Drzal 1964), for example, in the formation of moonmilk (Caumartin and Renault 1958).

IV

MINOR SOLUTION SCULPTURE

Weathering etches minor forms of diverse nature in many rocks, but the greatest variety is undoubtedly found in karst because of the susceptibility of karst rocks to solution. In German, *Karren* has come to be used as a comprehensive term for small-scale solutional sculpture (Bögli 1960) and for some French investigators *lapiez* has the same wide sense. Though there are transitional and compound forms, the multiplicity of distinct types has required further terminology such as Bögli has employed. Unfortunately in English there is no general term equivalent to *Karren,* and confusion and inadequacy rule amongst specific terms. A partly new nomenclature will be adopted here since to use a whole foreign terminology untranslated is neither euphonious nor readily assimilable.

FACTORS AFFECTING MINOR SOLUTION SCULPTURE

Many factors interact to control the nature and pattern of small-scale sculpture on karst rocks, and consequently that sculpture is far from being properly understood. Perhaps the most important control is the presence or absence of cover (soil, plant litter, superficial deposits, and vegetation itself) because the conditions affecting solution are very different in these two circumstances. Nearly all of the limestone of the Peak District of Derbyshire, England, is covered chiefly by aeolian mantles (Pigott 1962), so there is virtually only sculpture at the subsoil interface. On the very similar Carboniferous Limestone of the nearby Craven district of Yorkshire, there is much more rock exposed and much more obvious solutional activity. This is partly due to the last glaciation stripping earlier weathering mantles from Craven whereas the Peak District was not overridden by this ice sheet (cf. a similar contrast within the Jura Mountains; Aubert 1969).

However, the country rock itself may inhibit the formation of many kinds of detailed sculpture. Thus the porous, weak Cretaceous Chalk of southeast England and northern France is unfavourable to nearly all such landforms and there is not such a sharp interface between rock and soil with it as with most limestones. The same effect is found from different causes on the contact-metamorphosed Silurian limestone of the eastern side of Cooleman Plain, New South Wales. It is a hard, coarsely crystalline rock and weathers to rounded forms in outcrop, closely resembling those of glacial abrasion. Joints generally find little surface expression and any protected recesses carry a calcite gravel skeletal soil. The readiness with which the rock disaggregates operates against solutional sculpture.

Rapid changes of lithology or closely spaced bedding, jointing, or cleavage planes can also prevent the development of many forms. Increasing percentage of impurities affects the nature of the forms that do develop, especially producing more rounded edges and ribs.

In cold climates, thaw-freeze replaces solutional sculpture altogether (Corbel 1954). Thus the marble of Mt Arthur, Nelson, New Zealand, has an interesting range of solutional forms on outcrops above the treeline, but in the final one to two hundred metres below the summit there are only angular corners and platy surfaces on bedrock, with active scree formation. Temperature also exerts an influence through its effect on the chemical reactions of the water films. A probable consequence is that many forms are much longer downslope in tropical climes than in temperate ones.

The nature of precipitation enters into the matter, certain forms being associated with snow cover, which must act rather like a temporary (but biologically sterile) soil. The amounts, duration, and intensity of rainfall critically affect solution of bare rock and, through soil water, of subsoil surfaces. The most obvious precipitation effect is the extreme one of aridity, which minimises and deflects solutional activity. Solutional forms are few, for example, in the Nullarbor Plain of Australia (Jennings 1967a). The dominant result may be one of case-hardening of the exposed rock surface through evaporation, leading to reprecipitation of much of the modest amount dissolved. Accompanying this crust formation there may be the development of tafoni as in the Hadhramaut, Arabia (Wissmann 1957). The complexity of the matter is illustrated by the fact that, in a comparatively humid climate in various areas of Canterbury, New Zealand, Oligocene limestone, chiefly a porous shelly and foraminiferal biomicrite, also weathers to a rounded, hardened outer surface, broken by large tafoni; this is in part a lithological effect.

Historical factors involving climatic change have been inadequately evaluated, but may be important in some areas. In the Limestone Ranges, West Kimberley, Australia (Jennings 1969), Devonian reef rocks exhibit a tremendous array of bare karst solutional sculpture, yet the rainfall is a modest 450-700 mm annually with evaporation of the order of 2500 mm. Though the minor forms are so sharp they must still be developing, it may be unrealistic to attribute them wholly to a short season of a few, if intense falls of rain. The question arises whether Pleistocene pluvials (of which there is as yet no local evidence) were responsible for their creation, with solution since then sufficient only to prevent their degradation by other processes.

A factor which explains many patterns of sculptural forms as a result of exposure of subsoil surfaces through soil erosion is the destruction of forests by man and his grazing animals, loss of plant litter by oxidation on exposure, and loss of soil and litter by accelerated erosion consequent on agricultural activities. These effects can be seen to be operative today in areas where deforestation for grazing has recently been carried out or is still going on, e.g. Mole Creek, Tasmania, and Te Kuiti, New Zealand. But they are startlingly obvious where dense gardening populations fell and burn tropical rainforest, e.g. in the Central Highlands of New Guinea. With shifting agriculture, fresh subsurface forms, starkly blanched, emerge from litter and soil season by season (Pl. 10). Once this process is seen in action, the reconstruction of prehistoric forest clearance's denudative effects on limestone surfaces by palynologists such as that by Oldfield (1960) for the karst areas around Morecambe Bay in northwest England carries complete conviction. There can be no doubt that very many *Karren* in central and western Europe must be regarded as stripped. This was realised earlier in the Mediterranean lands. Indeed the certainty of this kind of history leads conversely to inferences about length and intensity of human occupation from

the distribution of sculptural forms. Within the Central Highlands
of Australian New Guinea, it has been postulated on other grounds
that the eastern parts were settled earlier than the western. That
at very least high density of population and intensity of land use
were achieved much sooner in the east than the west is suggested
by the presence of bare karst sculptural forms in the east and
their absence from the west except on sheer precipices. There
have been at least 2000 years of agriculture during which rock outcrops artificially exposed to the elements in the east could be so
modified.

Three groups of factors are therefore responsible for a very
complex overall *Karren* pattern: the *passive* factors of rock
petrology, porosity, bedding and jointing, and presence or absence
of covers; the *active* factors of quantities, temperature and acidity
of rain and soil water, plant growth, glacial preparation; and the
historical factors of changes over time of the preceding controls.

TYPES OF MINOR SOLUTION FORMS ON LIMESTONE

Some of the more common forms of small-scale solutional sculpture on limestone will now be· described. The classification is
based on that of Bögli (1960, 1961a) with some omissions and
additions. Littoral forms will not be discussed and related cave
features will be mentioned in Chapter VII.

I. *Forms developed on bare karst*

These forms develop with free movement of water uninterrupted
by mosses, liverworts, soil or sediment. Additionally it seems
necessary that the surface should not be so overhung by taller
vegetation that the fall of rain is significantly affected.

A. *With areal wetting.* The simplest effect of rain falling on
bare rocks is to produce small pitting, each *rainpit* being usually
less than 3 cm across and 2 cm deep. Rainpits form on gentle
surfaces rather than steep ones and may be separated by original
surface or become so close-set as to have only sharp rims between
them. Then the surface has an irregular, carious appearance. In
the tropics, solution flutes (see below) are sometimes interrupted
at intervals by rainpits along their length.

4 Horizontal solution ripples in Silurian limestone, Cooleman Plain, N.S.W. Scale 6 inches long.

Another very simple form is solution rippling (Pl. 4). This is found on steep (20-30°) to vertical surfaces and consists of practically horizontal, very shallow *solution ripples*, extending laterally over tens of cm but with each ripple 2-3 cm high. The edges between them are sub-rounded and the surfaces are much smoother than with pitting. These are well developed on Cooleman Plain, N.S.W. and the 'crinkled' weathering of Craven, England (Sweeting 1966, p. 197 and Pl. 22) seems to be related. The solution ripples of Wall and Wilford (1966) found in the tropical humid karst of Sarawak differ in that they occur on underhangs and are narrower and deeper with sharp ribs between. Moreover the water trickling over them is not fresh rainwater but has already had much contact with rock, humus, and soil. However, both are wavelike forms transverse to downward water movement under gravity, implying a definite rhythm in flow or periodicity of chemical reaction about which nothing is known at present.

More widespread and more striking are solution flutes (Ger. *Rillenkarren*), again found on fairly steep to vertical surfaces

5 *Solution flutes and bevel in Devonian limestone, Wee Jasper, N.S.W.*

(Pl. 5). These are longitudinal hollows, running in sets straight down the steepest inclination with sharp ribs between. There is a strong modality in cross-section, with a width of 2-4 cm and depth of 1-2 cm maintained uniformly. Length is much more variable, measured in tens of cm in temperate latitudes but often in metres in hotter climates. Their length and frequency are also related directly to rainfall. Where flutes develop on opposed sides of a block, a serrated crest results with a herring-bone pattern seen from above.

Rippling may occur in combination with fluting to give a netted appearance to the rock.

More frequently associated with solution flutes on moderately sloping surfaces are *solution bevels* (Ger. *Ausgleichsflächen;* Pl. 5). These are very smooth, nearly flat elements, forming micro-treads backed by steeper, fluted risers, which may curve in arcs round the upper margin of the bevel. In snow climates, the risers at the back of bevels are steeper and smoother; when these are arcuate in plan, the result is the *solution funnel step* (Ger. *Trittkarren, Trichterkarren;* Pl. 6). On vertical surfaces, beyond a certain length the ribs between two or three neighbouring solution flutes die out, leaving *rain solution runnels* (Ger. *Regenrinnenkarren*) of the second type of Bögli (1960) (Pl. 33). These forms are about two or three times as big as the solution flutes themselves, but like them retain the same cross-section throughout their length.

B. *With concentrations of runoff. Solution runnels* proper (Ger. *Rinnenkarren*) differ from the forms just described especially in that they increase in width and depth downstream through gathering volumes of runoff water. The ribs between neighbouring ones though substantial may be rather sharp. Bögli states that runnels are the product of slower solution than produced the features already described. This solution is dependent on diffusion of CO_2

6 *Shallow solution funnel steps in Palaeozoic marble, Mt Arthur, New Zealand. In rear, grikes, rain solution runnels, and solution ripples.*

from the air into the water, and is the stage at which large amounts of limestone are dissolved.

Not only does water collected from areas of bare rock act in this way (it is common for sets of solution flutes to feed into solution runnels), but so also does water running from moss polsters, from snow patches and from soil and humus covers; furthermore seepage water re-emerging from bedding planes and joints may participate. So it operates in many different circumstances. We can recognise therefore related forms such as *meandering runnels* (Ger. *Mäanderkarren*) which wind over flattish surfaces (Pl. 7), and *wall solution runnels* (Ger. *Wandkarren*) which are conversely very straight runnels due to water pouring down vertical faces. The latter are deeper and less regular than rain solution runnels, nor are they separated by narrow, sharp ribs.

Most important and widespread are *grikes* (Amer. solution slots; Ger. *Kluftkarren*) which are solution widened joints or cleavage planes (Pl. 9). These inherent planes of weakness canalise flow and so promote their widening. On flat surfaces especially, they may form by the merging of a series of vertical holes, lens-like in horizontal section, arranged along the joints. Grikes can be extremely straight, deep and long, and often occur in networks. Enlargement and rounding may take place at joint intersections to produce cylindrical pits several metres deep, known as *karst wells*. These are related both to solution pipes (p. 50) and potholes (p. 151).

When rocks are steeply dipping or vertically disposed, bedding planes are likely to be enlarged in the same way; they may be termed *bedding grikes* (Ger. *Schichtfugenkarren*). The strike ribs left between them may, however, differ from the residual blocks

7 Meandering solution runnels in calcreted Tertiary limestone, Nullarbor Plain, Western Australia

left between joint grikes. These ribs often break up into pinnacles (Ger. *Spitzkarren*) and beehives decorated by solution flutes.

When grikes develop in horizontal, thin-bedded limestone, the uppermost bedding planes are also likely to be opened up by seepage. This results in freeing the intervening tabular blocks known as *clints* (Pl. 9). These get shifted about as solution beneath disturbs their equilibrium and eventually they break up into irregular smaller fragments strewn about. *Shillow*, a term from the north of England, deserves employment for this latter condition (Pl. 33). It approximates to the German *Trümmerkarren* and *Scherbenkarst*, though these imply the likelihood of frost wedging. It would be better to eschew this connotation for the general term and use a qualifying adjective defining the process where appropriate.

II. *Forms developed on partly covered karst*

Some areas are patchily covered by soil, sediment, humus, moss polsters, or plant litter. It is important to bear in mind that this patchiness may be a result either of removal of a formerly more continuous cover or of progress in formation of such cover. Certain forms are characteristic of these conditions because water is retained in the patches of cover which usually supply biogenic CO_2 and, spongelike, permit sustained and substantial solution over the whole interface. Thus *grikes* commonly have soil or humus along their bottoms, with plants growing in the protected habitat.

To be classed here although they themselves may provide the only cover, are *solution pans* (Amer. tinajitas, etched potholes; Slav. *Kamenitsa*). These are basin or dish shaped depressions, usually covered by a thin layer of algal remains, silt, clay, or gravel. The organic matter supplies additional CO_2 and the fine clastics may seal off the flattest, lowest parts of the bottom from corrosion. In this way corrosion is concentrated on the steep to overhanging sides of the pan and extends the flat bottom. Indeed the most favourable locus of solution is the rock-air-water line; here attack will be most continuous through the renewal of CO_2 precisely where sealing is least likely. However, it must be noted that Williams (1968) found higher acidity in the organic and inorganic material at the bottom of a pan than in the pool of water itself. Centripetal patterns of solution flutes may surround the pan above the highest water levels. Under forest, solution pans can be preferentially enlarged through dripping from tree branches; here they are not accompanied by flutes. Related but deeper forms have been called solution cups by Zotov (1941) who infers they have formed with mosses in them since deforestation last century. However, such forms are found beneath forest and the problematic cases are those without any trace of organic or inorganic matter on their bottoms.

Closely related forms are *undercut solution runnels* (Ger. *Hohlkarren*) with bag-like cross-sections. These are attributed by Bögli to the filling of solution runnels on gently sloping surfaces with humus, soil or litter, whereby the sides and bottom are persistently wetted by water rich in biogenic CO_2 and thereby recessed with respect to their uncovered top.

Solution notches (Ger. *Korrosionskehlen*) are larger but usually still minor landforms due to particularly active solution in the top few decimetres of the interface between soil and projecting rock. This results in curved incuts at soil level up to a metre high which may run for many metres laterally. They are best developed in tropical climates (Pl. 8). Lowering of the soil surface can expose the whole of such features as in Pl. 8 though this event is not always a result of anthropogenic soil erosion.

Swamp slots (Wilford and Wall 1965) may be restricted to humid tropical conditions. They are horizontal smooth grooves with very flat roofs, a few centimetres to a few decimetres in height and up to about a metre or two deep. They can be found with swamp water and organic muck in them at the foot of cliffs or may occur in series high and dry above this level. In the latter case they are inferred to have formed when the swamp surface lay higher.

III. *Forms developed on covered karst*

Soil or sediment bears like an acidulated sponge on the underlying surface and produces its own array of forms. These will only be seen in excavations unless drastic denudation has removed unconsolidated inorganic and organic materials which have acted in this way (Pl. 1). However, in special circumstances where calcareous moraine covers the limestone, subsoil sculpture may be prevented and glacial striae may even survive, as Williams

8 *Solution notch in Eocene limestone, Muriraga, Central Highlands, New Guinea. Soil removal in abandoned gardens, exposing smooth incurve formed beneath soil cover.*

(1966a) found in Ireland. In such cases the solvent capacity of percolating water has been entirely neutralised within the drift. Figure 8 shows how solution forms gradually disappear when traced from a bare limestone pavement to a glacial moraine cover gradually exposing rock. Ice erosion has removed former solution sculpture and subsequently moraine has protected it. Aubert (1969) describes a similar case in the Jura.

Movement of soil water is not very free and so the predominant effect is for movement to be strongly guided by the rock structure. Therefore *grikes* are produced very effectively in these circumstances. Very many of the grikes of Craven, England, where the term originates, evolved with such fill but are now devoid of soil or humus, the change being due to interference with the geomorphic system by farmers from Neolithic times onwards (Pl. 9). In the United States, the term 'cutters' is commonly applied to soil-filled grikes exposed in quarries (Howard 1963). Below a certain depth, water movement is very slow indeed and the microstructures of the rock are etched out on the walls of open joints and bedding planes.

Within a metre or so of the surface, however, soil water movement is more strongly influenced by gravity and probably the rate of solution is faster. In these circumstances the minutiae of rock structure are overborne by corrosion and particularly smooth surfaces are developed. If any hollow begins to develop, it gathers

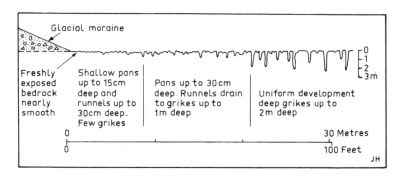

8 Relationship of solution sculpture to moraine cover in Clare, Ireland. After Williams 1966a.

more water to itself and promotes its own growth (Tricart and Silva 1960). Thus gravity controlled longitudinal hollows develop beneath the covers; they resemble solution runnels except that the ribs between them get rounded by the omnipresent moist blanket. These are the *Rundkarren* of the German literature and may be called *rounded solution runnels.* Again they are more familiar to us when stripped by subsequent erosion. The rounded runnels which form dendritic patterns on the nearly horizontal surface of clints in Craven (Pl. 9) are of this nature since they can be traced beneath nearby glacial drift (Sweeting 1966). However, the possibility of forest extending in recent times over bare rock surfaces on which ordinary solution runnels have developed previously must be borne in mind; these may have had their keels rounded and have been converted to the form characteristic beneath complete covers under forest litter. This sequence could clearly occur during Postglacial climatic amelioration following glacial erosion, and Bögli maintains this has happened around the Bödmeren area, Muototal, Switzerland.

In this zone close to the soil surface much less systematised smoothed surfaces also develop rounded dimpling and deeper hollowing which may penetrate as tubes through projections. This is the cavernous subsoil weathering of the German literature, *kavornosen Karren.* It is well exhibited in Highlands New Guinea gardens on limestone (Pl. 10).

Subsoil features which may become very large and pass over into major forms are vertical *solution pipes* (Ger. *geologischen Orgeln*). However, many of these are quite small cylindrical or

9 *Limestone pavement, Craven, England. Former soil cover eroded to expose grikes and rounded solution runnels. First row of clints about 1 metre wide.*

conical holes, a few decimetres in diameter and less than a metre in length, particularly when they develop in series along enlarged joint planes (Fig. 9). In the Chalk of northwest Europe they can develop in more isolated fashion unrelated to joints (Kirkaldy 1950). In aeolian calcarenites in southwestern Australia sections through earth filled pipes show that solution is not the only process involved in this special context of induration of dune sands into limestone (Jennings 1968). The pipes have a shell of calcrete around them through reprecipitation of calcite; the developing pipes are centres of induration. Roots are common in earth fills and pass down beyond into the calcarenite. Again there is an autocatalytic or positive feedback relationship. Root exudates and root respiration help the solutional deepening, and enlargement of the pipe promotes plant growth. Taproots of trees may completely fill the lower parts of pipes. On emerged coral reefs pipe development may be promoted by guano which corrodes the limestone.

Root activity is not restricted to pipes. Wall and Wilford (1966) have described root grooves of varying size and pattern in joint planes in Sarawak karst. In the Highlands of New Guinea roots can penetrate massive limestone without the help of planes of weakness and even vertical faces carry vegetation which riddles the surface with grooves and holes. Tricart and Silva (1960) and

Jennings and Sweeting (1963) draw attention to the importance of tree roots in seasonally wet tropical climate in Bahia, Brazil and northwest Australia respectively. Even in the dry climate of the Nullarbor Plain, Australia, intense perforation of the limestone close to the surface is, according to Lowry (1969), also due to tree roots.

MINOR SOLUTION FEATURES ON OTHER KARST ROCKS

The processes producing small solution features in dolomite are fundamentally similar to those in limestone (Chapter III). A detailed study of the features characteristic of dolomite outcrops of the Niagara escarpment has been made by Pluhar and Ford (1970). The dominant forms are controlled by rock structure, chiefly grikes along joints, often rectangular in pattern. Gravity controlled forms are few though there are some solution runnels down the vertical sides of grikes. There are also horizontal grooves along these sides but these are controlled by stylolite seams. The

10 Dimpled and pocketed, yet smoothed solution forms in process of exposure by removal of soil in gardening at Mainomo, Central Highlands, New Guinea

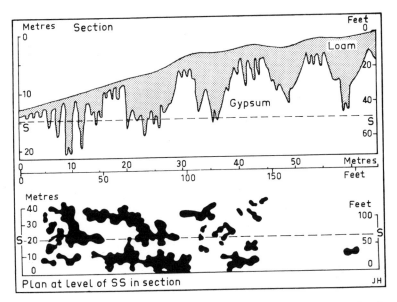

9 *Solution pipes in gypsum at Walkenried, Harz, Germany. After Penck 1924.*

grooves are developed in zones of coarser crystal size and higher porosity due to weathering between the stylolite seams, which project as blunt round ribs. Variations in porosity also seem to localise fairly numerous solution pits (deep solution pans).

In the seasonally dry tropics of northern Australia, dolomites as at Camooweal, Queensland, also show the dominance of structural control, grikes and clints being characteristic. Their bare surfaces show no flutes or runnels, having an overall carious aspect through tiny pitting.

Sculpture on non-carbonate karst rocks such as gypsum, anhydrite, and rocksalt is governed by physical solution without the interaction of solvents such as carbonic acid (Priesnitz 1969). In humid climate, the sheer rate of solution and mechanical weakness of gypsum produce unstable slopes in which *Karren* have little chance to develop, as below Pointe de Jacquette, Bas Queyras, in the French Alps (Pl. 2). In drier climate, this is not so and, even on the more soluble rocksalt, surfaces become intricately cut up with flutes and runnels as at Slanić in Romania (Krejčí-Graf 1935) and at Djelfa in Algeria (Würm 1953).

ASSEMBLAGES OF MINOR SOLUTION FEATURES

Where extensive areas of bare or partly covered karst occur, various types of sculptural form may be found in association, sometimes systematically. Their relationships may permit their order of development to be determined (Williams 1966a): Fig. 10 shows how the relative ages of runnels and grikes which intersect one another can be inferred.

In Craven, England, the patterns of associated grikes and clints vary with the spacing of joints, the thickness of beds and their dip (Sweeting 1966). With moderately thick beds and widely spaced joints, clints are better developed and are more likely to carry solution pans and rounded solution runnels than with very thin beds and very close jointing. With increasing dip, pans disappear and runnels adopt parallel courses down the clints, rather than dendritic ones.

RATES OF SOLUTION IN SUPERFICIAL ZONES

Study of the rates at which solutional sculpturing goes on has been a major activity in quantitative karst geomorphology during the last two decades. There are many practical problems such as that of measuring what goes on at the soil/rock interface. Moreover the circumstances are so full of variety that results are often not properly comparable with one another.

In 1947 Sweeting (1966) stripped moraine from a glaciated pavement in Craven, England and by 1960 the glacial striae exposed had been removed by corrosion, with a surface lowering as much as 3-5 cm in places. From another area, peat was removed and peaty water directed to flow on to this stripped rock surface; this cut runnels 7-15 cm deep in the same period. Experiments of this nature, though valuable, upset natural conditions and results obtained may not have a great deal of application elsewhere. Installation of stainless steel pegs as reference points causes much less disturbance (Hodgkin 1964) and future measurements with their help should have more significance.

Measurements of solutes in waters forming these solutional features today are not readily interpreted because of varying length and intensity of precipitation, varying length, rate and duration of surface flow, thickness of film, and presence of ions complicating the main solution process. Sampling water flowing over limestone

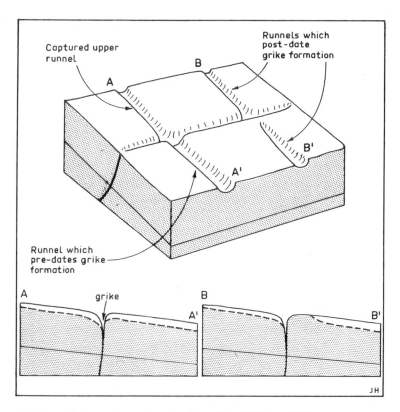

10 *Time relations of runnels and grikes. After Williams 1966a.*

on the Dachstein plateau, Austria, Bauer (1964) found that the concentration of $CaCO_3$ varied from a maximum of 40 mg/l at the start of a rainstorm to a minimum of 13 mg/l shortly after the greatest rate of rainfall when rate of flow outbalanced the solution rate, rising again to 25 mg/l during the dying phase. He also compared the concentrations achieved in flows down a smooth slope (14-63 mg/l), a shallow runnel (15-76 mg/l), and a deep one (30-96 mg/l). Converting the mean values to long-term rates of surface removal by applying them to the mean annual rainfall, he arrived at rates of horizontal surface lowering of 0·9 cm/1000 years, 1·0 cm/1000 y and 1·3 cm/1000 y respectively. This shows how runnels promote their own development (autocatalysis).

At the same time Bauer's figures indicate that small numbers of somewhat haphazard samples from different kinds of karst are not likely to provide very meaningful comparisons. Table 3 sets out some of the available data. Williams's large number of samples from solution pans from Clare, Ireland, provides a reliable mean of 66 mg/l which for the prevailing temperature and normal atmospheric P_{CO_2} corresponds with the theoretical saturation equilibrium. Because of very free interchange with the atmosphere, the algae in the pools can only increase the speed with which saturation is reached and do not enable a greater equilibrium concentration to be achieved here. The moss polster and the soil/rock interface samples have generally higher concentrations than the bare rock samples, though there is no significant difference in the Yugoslavian data and there is a reverse relationship in the Lapland case.

Because the soil provides the bulk of the biogenic CO_2, a great deal of solution takes place inside the soil as long as carbonate fragments remain and at the surface of contact between the soil and the bedrock in place. Measurements of $CaCO_3$ concentrations in the Jura have shown that this superficial solution is equivalent to the removal of a layer of limestone 0·05 mm thick each year and it is thought that this is about 60 per cent of the total amount going into solution in the area (Aubert 1969).

Investigators have generally found their highest concentrations of $CaCO_3$ in waters passing through unconsolidated deposits containing many limestone clasts. Thus on the Dachstein plateau Bauer (1964) determined the rate of calcium carbonate loss from a glacial moraine through the amount removed to create small depressions in its surface overlying tongues of decalcification (Fig. 11). The rate of loss of rock equivalent was 3·6 cm/1000 y compared with 2·8 cm at the forest humus-rock interface and the much lower rates from bare surfaces mentioned earlier.

This last determination avoids the danger in extrapolating backwards in time from observations of the conditions at the present moment. Thus Sweeting (1966) points to the great reactiveness of smoke from industrial areas of northeast England which reaches limestone areas of Craven, for example, and it is well established that climate changes significantly over quite short periods of time.

Therefore a great deal of interest attaches to surface losses attested by the pedestals (Fig. 12) which have formed beneath clints and glacial erratics on limestone pavements eroded by ice of the last Pleistocene glaciation (Table 4).

TABLE 3 **Carbonate concentrations of superficial waters**

Karst area	Köppen climatic type	Mean CaCO₃ concentration in mg/l from			
		Bare rock slope	Pan pool	Moss on rock	Soil/rock interface
Somerset I., N. Canada	ET	46(29)Sm			
Lapland, Norway	Dfc	71(6) M		49(3)M	
Bisistal, Switzerland	Dfb	21(6) B		39 B	
Dachstein, Austria	Dfb	34 Ba			100(1) Ba
Jura, France	Cfb	71(26)A			122(38)A
Sarthe, France	Cfb			50(1)M	106(5) M
Craven, England	Cfb		131(3) S		
Clare, Ireland	Cfb	37(4) M	66(118)W		
S.E. Australia	Cfb		81(3) J/S		
Nullarbor Plain, Australia	Bsh		75(3) J		
W. Kimberley, Australia	Bshw		109(3) J&S		
Puerto Rico	Amw				60(5) M
Jamaica (dry season)	Aw				55(2) M
(wet season)					68(14)M
Cuba	Amw	90(1) L			115(3) L
Indonesia	Afi		150 Bl		

(n) = number of samples determined.
Sources: Sm = Smith 1969; M = Muxart and others 1968; B = Bögli 1960; Ba = Bauer 1964; S = Sweeting 1966; W = Williams 1968; A = Aubert 1969; J/S = Jennings and Sweeting (unpublished data); J = Jennings (unpublished data); J & S = Jennings and Sweeting 1963; L = Lehmann and others 1956; Bl = Balázs 1968.

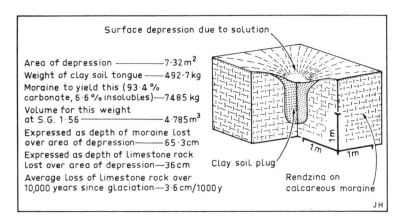

Surface depression due to solution

Area of depression ———————7·32 m²
Weight of clay soil tongue ——492·7 kg
Moraine to yield this (93·4 %
carbonate, 6·6 % insolubles)—7485 kg
Volume for this weight
at S.G. 1·56 ———————————4·785 m³
Expressed as depth of moraine lost
over area of depression———65·3 cm
Expressed as depth of limestone rock
lost over area of depression—36 cm   Clay soil plug
Average loss of limestone rock over
10,000 years since glaciation—3·6 cm/1000 y

Rendzina on
calcareous moraine

JH

*11 Determination of limestone solution from pit in rendzina on calcareous
moraine in the Austrian Alps. After Bauer 1964.*

The good agreement between the results achieved by the two
methods is therefore surprising but very useful. The conclusion
follows that very many surface solution features have formed in
the Holocene. This agrees with the observations that they are
commonly found in areas which were strongly abraded and plucked

*12 Glacial erratic of Silurian sandstone on a Carboniferous limestone
pedestal due to Postglacial solution, Norber Crags, Craven, England.
Sketch by C. D. Ollier.*

TABLE 4 **Surface lowering rates from limestone pedestals**

Area	Average ht of pedestals cm	Time since ice disappearance years B.P.	Rate of surface lowering		Proportion of total limestone removal from area
			from pedestals cm/1000 y	from runoff & precipitation cm/1000 y	
Mären Mts, Switzerland (Bögli 1961a)	15	10,000	1·5	1·51	?
Craven, England (Sweeting 1966)	50	12,000	4·2	4·0	50%
Clare-Galway, Ireland (Williams 1966a)	15	15,000	1·0	0·5–1·5	25%

by glacier ice in the final phase of the last glaciation (Williams 1966a; Haserodt 1969). Nevertheless it must not be assumed that all minor solution features, especially deep grikes, formed in such a short period.

V

DRAINAGE

Surface drainage in karst is liable to be intermittent, disrupted, widely spaced, or absent. Marked permeability has enabled underground drainage to take over the task of moving water to the sea or to surrounding country on other rocks to varying degrees in space and time (Pl. 11). In this chapter the passage of water through karst will be discussed separately from the resulting landforms as far as possible so that the functioning of karst hydrologic systems will stand out more clearly.

INFILTRATION, OVERLAND FLOW, AND THROUGHFLOW

Karst is marked by rapid and substantial infiltration into rock outcrops and soil, restricted overland flow which rarely reaches stream channels, and modest throughflow in the sense of lateral movement through soil pore space, though these generalisations need some elaboration and qualification.

Falling on karst rock outcrops, rainwater usually flows over the surface very short distances before infiltrating. The more intense the rainfall and the steeper the surface the longer will overland flow be, as witness the long solution flutes and runnels on vertical faces in the perennially or seasonally humid tropics. The coarse porosity of emerged coral reefs usually causes immediate absorption so that lack of surface water can be extreme even with mean annual precipitations of 2000 mm or more. Rapid infiltration is also characteristic of chalk despite its micritic texture and very fine intergranular porosity. With compact rocks lacking intergranular porosity, water flows over the surface till it encounters planes of weakness in them. The frequency, openness, and continuity of these planes controls infiltration. Aubert (1969) contrasts

11 River Angabara, Central Highlands, New Guinea, penetrating through Miocene limestone strike ridge. Flat-floored closed depression beyond the ridge is partly developed on impure limestone.

the numbers and openness of joints, particularly strike joints, along anticlines in the Jura Mountains with their paucity and tightness in the synclines; very fine joints, especially characteristic of argillaceous limestones, hinder infiltration as a result of surface tension and blockage by weathering products and precipitates. Joints wide enough to permit infiltration are liable to be enlarged by solution so that any initial permeability accentuates itself. Close fields of grikes bring about virtually immediate loss of rain underground. In thick beds lacking porosity, solution pans and larger solution hollows, such as the rockholes of the Nullarbor Plain (Jennings 1967c), can hold water until it is entirely lost by evaporation.

With covered karst, the nature of the soil or superficial deposits is critical. Many residual soils on karst rocks allow high infiltration rates. This is true of rendzinas, common in mid-latitude karsts; these soils are dark, alkaline, shallow loamy soils, with crumb structure and including residual rock fragments. Such soils also permit throughflow but this will only be significant on steep slopes since infiltration water tends to pass rapidly into the underlying bedrock with but modest lateral movement through the soil pores or at the base of the soil.

However, other residual and transported soils in karst behave very differently. Prolonged leaching of the weathering mantle can result in acid, dense clay soils which consist of clay mineral and iron sesquioxide residues from the bedrock. Some at least of the 'terra rossa' commonly found in Mediterranean karstlands is of this nature. Soils like these restrain infiltration, especially when they are cleared of forest, grazed or cultivated; moreover they tend to block joints in the underlying rock. Longer overland flows prior to infiltration or to feeding surface streams result. Where the karst is only partly covered, less permeable soils of this kind generally occupy hollows and deflect water laterally towards outcrops. Heavy clay soils act as seals in the bottoms of depressions; hydromorphic soils, swamps, and even standing water may develop.

Superficial deposits are as various as the soils. Colluvia derived from limestone or dolomite are likely to include much coarse material and be very permeable, promoting both infiltration and throughflow on the appreciable slopes associated with them. Whether originating within or without the karst, alluvia will chiefly vary according to texture, the finer the texture the less the infiltration and throughflow and the greater the overland flow. Glacial moraine derived from the karst itself is commonly permeable and overland flow modest over it. Loess covers behave similarly and so also those of volcanic ash until pedogenesis in humid climates has made substantial progress, by which time they are also likely to have been stripped from higher relief and concentrated in the lower parts of the landscape. The bauxitic clays in the Jamaican karst cockpits, of low infiltration capacity, may be of this latter origin. This list could be made very long and it is important to recognise that there can be sharp variations in the relative importance of infiltration and runoff both within and between karst regions.

Vegetation also has a great influence on infiltration, largely through varying loss by transpiration. Holmes and Colville (1970 a, b) have shown that on the same Gambier Limestone and the same soil in southeastern South Australia, pine forest causes twice as much loss to the atmosphere as grassland and this reduces infiltration into the limestone to nil.

Surface streams are usually few in karst. In the whole Dinaric karst of Yugoslavia, only four rivers cross it to reach the sea—the Krka, Cetina, Morača, and Neretva—and of these only the last crosses the whole width of the main limestone belt. One-quarter of the whole area drains directly into the sea without surface stream flow at all (Gams 1969). In semiarid and arid karst, rivers are completely absent and this is rarely true of desert country on impermeable rocks. The Nullarbor Plain, Australia, a free karst like the Dinaric, feeds all its meagre water underground to the ocean.

In very humid climates or on less pure karst rocks, more elaborate river patterns are found, but generally the drainage density is less than on other rocks in the same conditions. With well developed karst this attribute is self-evident but in other circumstances it may need careful demonstration. Such an instance is provided by Miller's analysis of drainage basin characteristics in three lithologies in the Clinch Mountains area of Virginia and Tennessee, employing Horton's measure of drainage density of length of channel per unit area of catchment (Miller 1953). Dolomite is the bedrock in the Copper Ridge area whereas thick, coarse sandstones alternate with shales in the neighbourhood of Blountville, with the Pennington district providing the greatest contrast with thin, fine-grained sandstones and siltstones interbedded with a dominant body of shale. All have similar local relief and dendritic drainage but streams are longer and basins larger in accordance with degree of perviousness (Table 5). As a result drainage density is least in Copper Ridge and greatest in Pennington in both first and second order streams. The differences between the dolomite area and the dominantly shale area are significant at the 95 per cent confidence level. The Blountville area with its thick pervious sandstones as well as thick shales is only significantly different from the Pennington shales in drainage density of first order streams.

The four rivers which cross the Dinaric karst are allogenic (allochthonous) in that their headwaters are on impervious rocks. They have practically no surface tributaries along their entire

TABLE 5 **Analysis of drainage densities on different rocks in Clinch Mountains Area, Virginia and Tennessee (Miller 1953)**

	First order streams				Second order streams			
	Km/km^2	Copper Ridge	Blount-ville	Pennington	Km/km^2	Copper Ridge	Blount-ville	Pennington
Copper Ridge (dolomite)	X̄ 5·58 δ 2·48		−	+	X̄ 5·21 δ 1·76		−	+
Blountville (shale and thick coarse sandstone)	X̄ 5·99 δ 2·20	−		+	X̄ 6·02 δ 1·25	−		−
Pennington (shale and thin interbedded fine sandstone and siltstone)	X̄ 7·42 δ 2·47	+	+		X̄ 6·70 δ 1·66	+	−	

X̄ mean length   δ standard deviation

Differences   + = statistically significant   − = not statistically significant lengths through the karst, though springs do supply them with water. Allogenic rivers are very common in karst regions. However, there are rivers which begin their courses on karst rocks, i.e. they are autogenic (autochthonous). They do so in springs, often large ones, so that autogenic karst rivers are liable to be born adult, as it were. The circumstances which give rise to springs are discussed below.

To cross karst on the surface, both kinds of river are frequently dependent on alluvium which seals off the permeable country rock. Some tropical karsts consist largely of broad, alluvial plains on limestone, with perennial rivers meandering over them as in the Kinta valley in Malaya. In Guadeloupe dendritic valley systems occupy much of the karst with intermittent surface streams along their bottoms where terra rossa has accumulated thickly (Lasserre 1954). Although the Central Lowland of Ireland is karstic, it has many surface streams and lakes; this is undoubtedly partly due to a mantle of Pleistocene glacial deposits. But it is probably also a consequence of preglacial reduction by solution of the limestone to a plain close to sea level on which residual red clay cover had developed (Williams 1970).

RIVER REGIMES

The more underground drainage participates in a hydrological system the more efficient it is and the more of the precipitation can act geomorphically. Rapid infiltration means that water escapes the heat, wind, and low relative humidities of the surface sooner and so evaporation is reduced. It is true that plant roots probably reach more deeply as another consequence. Tree roots 20-30 m underground are common; in Lake Cave, southwestern Australia, living roots occur about 60 m down. Nevertheless biological productivity is not generally higher in karst than elsewhere, and is often less, so that increased transpiration does not balance reduced evaporation. Because of this, limestone areas discharge a higher proportion of the water they receive as precipitation than does impervious terrain. Pardé (1965) estimates that evapotranspirational loss, determined by the precipitation-discharge deficit, in the Nera, Aniene, and Pescara headwaters in the calcareous Apennines is about 500 mm whereas impervious catchments in these relief and climatic conditions lose 600 mm. This represents a 15 per cent saving. Equivalent figures for the Tarn in the Grandes Causses, France, are about 300-310 and 425-450 mm respectively, making it even more efficient with a saving of 30 per cent.

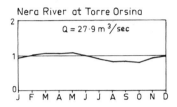
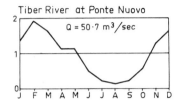

13 *Graphs of mean monthly discharges as fractions of mean annual discharge of a karst river (River Nera) and a normal river (River Tiber). After Pardé 1965.*

The water storage capacity of karst also affects river regimes and so modifies their geomorphic behaviour. Pardé (1965) estimates the percentage of voids at 10-20 per cent for small karst masses but at only 3-5 per cent for whole plateaux. However, the voids are virtually all effective in terms of permeability and storage. The result is that karst rivers sustain higher base flows longer and have only moderate flood peaks compared with rivers on impervious country. In Italy, for example, the mean monthly discharges of the River Nera at Torre Orsina only vary between 0·83 and 1·06 of the mean annual discharge, whereas the Tiber at Ponte Nuovo has corresponding figures of 0·18 and 1·95 (Fig. 13). The extreme minimum discharge of the Nera of 0·35 compares very favourably with the 0·07 of the Tiber. The maximum flood recorded on the Nera is only about 500 m³/sec but 3000-4000 m³/sec could be expected if it were on impervious terrain like the Tiber. White and Reich (1970) have similarly shown that mean annual flood is low in carbonate basins compared with that in basins without karst rocks with data from Pennsylvania.

This kind of flow regime might be expected to minimise river erosion in karst since it is accepted that rivers perform most of their work during floods, whether these are exceptional in magnitude or more numerous ones of lesser magnitude. Diminishment of power must apply to mechanical action by karst rivers. However, corrosion is much more important than corrasion in this context. Correlation of discharge and carbonate solute concentration in karst rivers generally shows an inverse relationship (Fig.

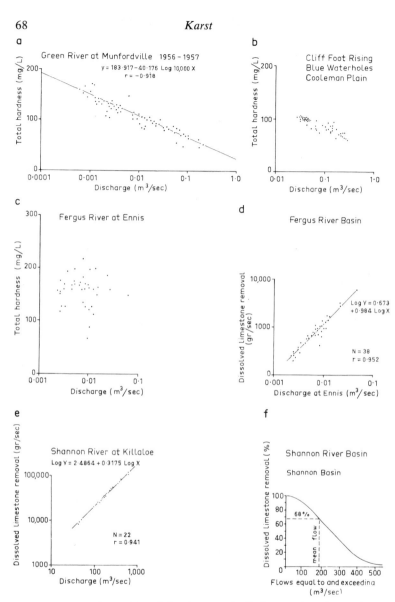

14 (a) *Inverse relationship of hardness and discharge of a karst river,*
     Green River, Kentucky. After Douglas 1968.

   (b) *Same relationship in a karst spring, Blue Waterholes, N.S.W.*

   (c) *Lack of correlation of hardness and discharge of a karst river,*
     Fergus River, Ireland. After Williams 1968.

   (d) *Direct relationship of limestone removal and discharge, Fergus*
     River. After Williams 1968.

   (e) *Same relationship in Shannon River, Ireland. After Williams 1970.*

   (f) *Relationship of limestone removal rates to flow frequencies, Shannon*
     River. After Williams 1970.

14a), though Douglas (1968) and Williams (1968) have shown that in the case of the Hull River, England, and the Fergus River, Ireland, there is no significant change of hardness with discharge (Fig. 14c). The River Thames near Oxford, England, even shows a direct relationship between carbonate hardness and volume of flow, but this is exceptional. Thus reductions of high stage in karst rivers and consequent diminishments of velocity are not such as to involve loss of solutional capacity as well; turbulence must generally be sufficient for the moving water to dissolve as much of the karst rock as other circumstances permit. It must be noted that the higher stages will still remove more limestone in unit time and for the Fergus, Williams (1968) found that in the winter month of December, with its higher discharge, four times more limestone was transported from the catchment than in either June or July (Fig. 14c). In the Shannon, also in the Central Lowland of Ireland, mean flow and lower stages account for about 68 per cent of the annual load of dissolved limestone (Fig. 14e, f; Williams 1970). Thus karst rivers can remain geomorphically effective with their moderated regimes because of their dominant solutional action. This would not be the case if corrasion was the chief way in which they deepened their valleys.

Nevertheless many limestone rivers may transport rather than corrode. Through his studies of water chemistry in the Fergus and Shannon catchments, Williams has shown that most solution takes place before water reaches the surface streams where indications of carbonate precipitation are the most prominent evidence. The constructional action of karst rivers is discussed in Chapter VI.

SINKING OF RIVERS

Rivers entering karst are liable to lose all or part of their drainage underground. Where the waters of a stream can be seen to enter a cave entrance, opening laterally (Pl. 12), or vertically from the surface channel, or to go into narrow fissures in the bedrock, the fact of karst loss is evident whatever the climate. In Slovenia there is the classic case of the Pivka River, which is barred by Sovič Hill, under which it passes quietly into the famous Postojna Cave. In Craven the stream from Malham Tarn sinks into fissures in its bed. In humid climates the loss is still evident, provided it is substantial or complete, even when it takes place gradually into

Karst

gravels on the bottoms and sides of river channels or in swampy areas, and thence into the karst rock beneath. The Takaka River in Nelson, New Zealand, is liable to disappear into its gravels whereas other rivers in the area in impervious catchments flow perennially over their gravel beds to the sea.

12 River Rak entering Tkalca Cave, Slovenia. Photo by P. W. Williams.

But in subhumid and semiarid climates rivers are intermittent in flow and may only reach part way along their courses as a normality through evaporation, percolation into gravels, and failing supply. Here additional loss into underlying pervious bedrock may not be readily discernible though it can be just as important in the hydrologic regime. Wee Jasper Creek in southern New South Wales feeds water into Dogleg Cave under Punchbowl Hill by ways too small to identify, let alone penetrate, where it flows along its gravel and bedrock bed along the flank of that limestone hill. However, this channel does not differ in general appearance from those of other streams in the area on impermeable rocks since all are intermittent in flow. Southwards some 55 km in the same drainage basin but at subalpine levels, the North Branch of Cave Creek is frequently dry over a substantial portion of its course across the limestone Cooleman Plain; this is remarkable because it is an area where water balances normally result in perennial drainage. Where only a modest part of the discharge of a stream is lost into a karst rock beneath gravels or swamps, it may be hard to detect without gauging even in a very humid climate.

Frequently a stream will have a series of sinking points or swallets along its course into which it loses successive fractions of its volume. The Manifold River in the English Peak District may fail to flow over 7 km of its bed between Wetton Mill and Ilam Hall in dry weather (Warwick 1953). The upper course of the Manifold runs over Namurian shales, then it flows on to Visean limestone with interbedded shales and finally encounters reef limestones (Fig. 15). There its discharge decreases through losses into a series of streamsinks. Some are shallow, bouldery depressions in the bed as at Wetton Mill itself; others are inflow cave entrances, e.g. Redhurst Swallet. With falling stage, the most downstream one becomes the limit of surface flow first, then successively each upstream one in turn until the Manifold gets no further than Wetton Mill itself. The continuity of surface flow is usually broken for 1-3 months each year. With rising stage, the Manifold exceeds the capacity of each successive sinking point in turn downstream and eventually has continuous flow over the surface once more.

Streamsinks can be ordered and analysed in much the same way as ordinary surface streams in the Horton-Strahler manner

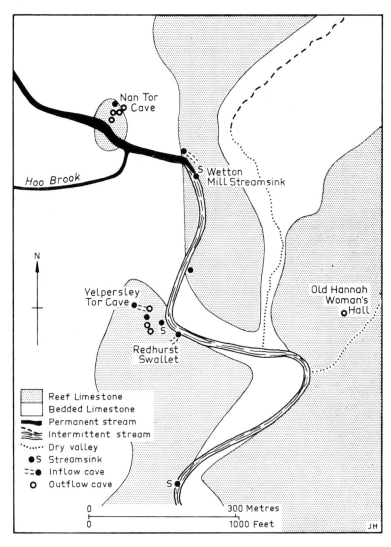

15 Successive streamsinks at Wetton Mill on River Manifold, Peak District, England. After Warwick 1953.

(Williams 1966b; Fig. 16). A first order stream has no tributaries, whereas a second order stream has tributaries but these are only first order streams. A third order stream can have first or second order streams as tributaries, and so on with higher orders. A streamsink takes on the order of the stream disappearing in it.

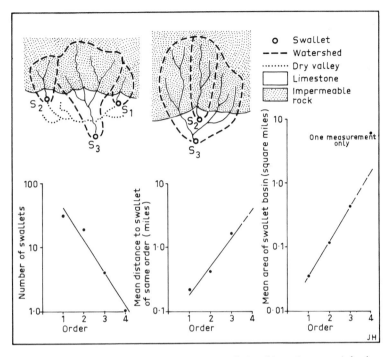

16 Method of ordering streamsinks. Some relationships of streamsinks for northern Ingleborough area, Craven, England. After Williams 1966b.

Williams modifies the ordering system in certain ways. Dry valleys on impervious rocks forming tributaries of sink systems are included in the reckoning since they are liable to have surface flows in wet seasons. On the other hand sinking streams fed from large springs (which are the resurgences of surface streams) are excluded from the analysis because they cannot be ordered properly. Determining the catchment area of a swallet is based on surface watersheds; this will be satisfactory as long as most of the catchment is on impervious rock and little on the karst where surface watersheds often do not correspond with underground ones. A higher order streamsink catchment may enclose those of lower order sinks; the extent of the latter must be excluded from the former's area.

Williams applied this system to the Ingleborough part of Craven and found that as with normal stream systems, the frequency of streamsinks varies inversely and geometrically with sink order, whilst the mean area of streamsink catchment varies directly and geometrically with sink order. Furthermore the mean distance apart of swallets of the same order increases geometrically with sink order. All these relationships are to be expected but they provide measures of organisation in streamsink systems. An index of streamsink density can be determined by dividing the total number of sinks by the area of karst concerned and provides one of the means of comparing the hydrology of different karsts.

There are some streamsinks which in time of flood may reverse their function and discharge water, acting like springs. This happens because the cave which a given stream is feeding also receives other feeders underground and on these occasions the combined volume is greater than the absorbing capacity of the cave farther in. The French name *estavelle* is commonly used for these alternating orifices but they are also well known by the Greek *katavothre*.

SPRINGS

In most kinds of terrain, throughflow in soil, waste mantles, and weathered rock gives rise to small and intermittent springs. In karst, springs are much larger and more frequently permanent because of greater infiltration of precipitation into karst rocks and because of the input of streamsinks fed from surrounding impervious rocks. A useful distinction can be made between *exsurgences* fed entirely by seepage waters from the karst and *resurgences* supplied by the sinking of surface streams (Trombe 1952). Resurgences are nearly always fed in part by seepage also so that there is every transition between the two. Moreover there are usually many springs of unknown provenance. Between Cape Naturaliste and Cape Leeuwin in southwestern Australia, dune limestone along the coast forms higher ground than the land just behind. Some drainage from the east enters the dune belt to feed resurgences near the coast; this is true of a spring west of Conolly Cave which is fed from Mammoth and Ruddock Caves, each receiving surface drainage. Farther south there are springs such as those at Cape Leeuwin on both sides of the dune belt which receive no surface inflow from without; these are exsurgences.

Karst springs are very varied. Some take the form of the open entrances to caves out of which streams flow under gravity (Fig. 17a, Pl. 13). Such are common in Craven because the valleys

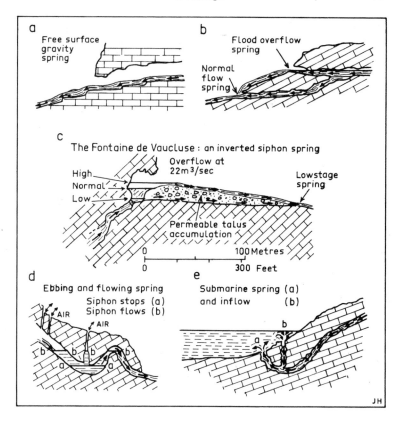

17 *Types of springs*
    (a) *Gravity-fed spring issuing from cave mouth.*
    (b) *Flood surplus spring.*
    (c) *Spring rising under hydrostatic pressure, Fountain of Vaucluse, Provence, France. After Martel and de Cousteau.*
    (d) *Ebbing and flowing spring. After Trombe 1952.*
    (e) *Submarine spring and inflow as in Sea of Argostoli, Kephallinia, Greece.*

there have cut down deeply into the limestone close to or into an underlying basement of impervious rocks. Some springs such as White Scar Cave and Austwick Beck Head are actually developed in the unconformity between the limestone and the Precambrian foundation. But others such as Clapham Beck Head, Douk Gill Cave, and Birkwith Cave lie within the limestone. Other springs descend similarly under gravity but emerge through the interstices in rubble which has fallen and accumulated over a cave entrance; this is the case with Cliff Cave Spring, Cooleman Plain.

13 River Lagaip emerging from a cave near Kepilam, Central Highlands, New Guinea

An alternative name for a spring is a rising and this name is literally earned by many karst springs which well upwards, quietly or with vigour, through open cavities or narrow fissures, gravels or swamps. There is quiet upwelling through gravels and in wide swamps at Kirk Göz in the north of the Plain of Pamphylia by Antalya in southern Turkey. These risings occur on the edge of an erosion surface against the Taurus Mountains, both in Cretaceous limestone. These are the waters of the Düdencay, which sinks again shortly for a further underground course of 12 km beneath Pleistocene travertine forming part of the same surface. The second emergence lies at the rear of a lower erosion surface cut into the travertine (Vaumas 1968). Set in a gravel plain, Piu Piu Spring near Takaka, New Zealand, is a perpetually 'boiling spring', rising with such force with a mean flow of 10.5 m^3/sec as to prevent divers from reaching the orifices of emergence which appear to be in white limestone (Pl. 14). It is probably a true 'Vauclusian spring' in that the water is rising under pressure up water-filled passages in bedrock. The Fountain of Vaucluse itself at the head of the River Sorgue east of Avignon in southeastern France has a bedrock passage rising at about $45°$ to a little round cave in

14 Piu Piu Spring near Takaka, New Zealand. 'Boiling spring' rising under pressure from limestone beneath alluvial plain.

which it forms a pool, extending out into the open at high levels when it overtops a shallow talus barrier (Fig. 17c). Normally it percolates through the talus above bedrock to emerge about 26 m lower down. Divers have descended over 100 m in the bedrock passage but have not been able to reach the descending part of the inverted siphon which provides the pressure head for this spring.

'Ebbing and flowing wells', intermittent springs with a regular period, such as that on Buckhaw Brow, Craven or the Arize River spring in the French Pyrenees, require a true siphon for their action (Fig. 17d). The water level in the system oscillates between a and b. When it builds up to b, the siphon begins to function and the level drops rapidly to a. There must be some connection to the atmosphere behind the siphon or else air pressure effects would intervene and the capacity of the siphon must be greater than the inflow of the cave stream, otherwise there would simply be a persistent small flow.

Frequently a persistent spring has associated with it a higher spring which only functions after heavy rains. The passage to the lower spring fills to capacity at this time, water backs up behind and there is overflow through a branch passage to the higher outlet (Fig. 17b). This is the relationship of Ingleborough Cave to Clapham Beck Head at Clapham, Craven, and on a larger scale of the Hölloch Cave to its perennial spring, Schleichende Brunnen, 100 m lower down in Muototal, Switzerland.

Many factors govern the location of springs. Underlying and interbedded impervious rocks cause springs to emerge from the base of the overlying karst rock. Impervious deposits such as glacial moraine banked against limestone or dolomite result sometimes in springs along their upper margin, and alluvial fills forming river terraces often have the same effect. Without the intervention of impervious materials, springs emerge along valley bottoms in karst or at sharp breaks of slope as with the case mentioned earlier. These are places where the saturated zone or rest levels in underground streams intersect the surface. Springs also occur high up valleysides within the body of karst rock without obvious topographic or lithologic cause; such is the case with the Golling Falls 100 m above the Salzach River in Western Austria. Here valley deepening has proceeded so fast that solutional enhancement of permeability has not been able to keep pace. The lower part of the limestone is still impervious and throws out the underground water. This appears to be the case also with the Efflux in Bungonia Gorge, N.S.W. Sea level is another important control of spring location, e.g. the large Ombla spring emerges at the shoreline near Dubrovnik on the Dinaric coast.

Nevertheless springs may emerge beneath lake or sea surface level where the karst rocks descend well below it (Fig. 17e). The Bourbioz spring emerges at −80 m in Lake Annecy in the French Alps. Vrana Lake in the island of Cres has sublacustrine springs which appear to be fed beneath the Strait of Kvarner from the Istrian plateau or from the Croatian plateau to the east. The writings of classical Greece and Rome have made famous the submarine springs of the Asia Minor coast, e.g. near Akcay in the Gulf of Edremit, of the Dinaric coast, e.g. near Rijeka, and of Greece itself. Submarine springs called 'posas' are also common in the sea around the limestone plateau of Yucatan.

Mistardis (1968) specifies a number of submarine springs emerging at depths of 30 to 40 m around southern Greece. Four hundred metres off the Kynourian coast with the high Parnon Mountain above, Anavalos spring opens on the bottom at −36 m but a diver has descended the spring itself a further 40 m so that the fresh water emerges at least about 75 m below sea level. Many writers including Mistardis attribute these deep submarine springs to cave development beneath the exposed sea floor during Pleistocene glaceoeustatic low sea levels. Theoretically deep phreatic solution (see p. 96) should also be capable of forming them beneath the sea.

Genetically related to submarine freshwater springs are submarine sinking points of sea water such as the famous whirlpool in the Sea of Argostoli in the Ionian island of Kephallinia (Fig. 17e). If a submarine spring has two openings, it is possible that the impetus of the freshwater current through the more direct passage may draw sea water down the other to mix with the spring water and emerge through the first outlet. Trombe (1952) has shown that the density differences between the columns of sea water and less saline mixed water can be sufficient to motor such a system on its own once it has started. Stringfield and Legrand (1969b) have used the same explanation for reversing flow of salt and fresh water between Tarpon Springs on the west coast of Florida and Lake Tarpon nearly 2 km inland.

Williams (1966b) has proposed several measures for the analysis of the geomorphic role of springs—the karst rising density, which is the number of springs divided by the area of the karst, and the rising coefficient, which is the ratio of the standard deviation of rising heights to the mean rising height expressed as a percentage. The rising coefficient may be useful in the interpretation of karst hydrology. Other related indices put forward are the vadose index (the difference between the mean heights of streamsinks and of risings respectively) and the sink/rising ratio. The mean shortest distance of underground flow is derived from the distances between each streamsink and its nearest rising below its own level. Insufficient application has been made of these indices yet to estimate their usefulness. The vadose index may in fact conceal some hydrostatic rise in the saturated zone if there are springs under pressure in the area.

REGIMES OF KARST SPRINGS

Most of the large springs of the world are karst springs though there are also very large concentrations of spring waters in basaltic lavas (see Ollier 1969). In terms of present knowledge, the group of springs at the source of the Manavgat east of Antalya in southern Turkey yields the most of all with an average discharge of 125-130 m³/sec. One spring alone at low stage has a flow of 40 m³/sec. Only 1000 km² of the southern flank of the limestone Taurus Mountains feed these springs which, however, tap also 25-30,000 km² of the northern slopes and the central plateau.

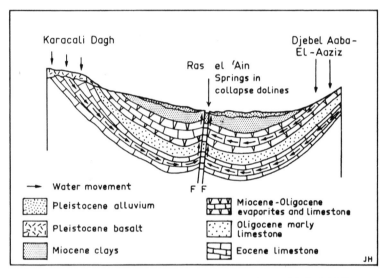

18 Artesian spring of Ras el 'Ain, Syria. After Burdon and Safadi 1963.

Just south of the Turkish-Syrian border, Ras el 'Ain with an average discharge of 38·7 m³/sec is the effective head of the Khabur River between the headwaters of the Tigris and the Euphrates. It is a complex of thirteen springs (Burdon and Safadi 1963) in two groups, each receiving true artesian water along a fault plane which taps aquifers in a synclinal structure (Fig. 18). The main aquifer comprises Eocene limestones with intakes on their own outcrop and on overlying basalts in the Turkish mountains to the north. Higher in the syncline are Miocene limestones and evaporites which probably furnish subordinate supplies from the south. The northern group of springs has a chemical content indicative of supply from limestones only whereas the southern group probably get water from the evaporites also. The springs appear as circular ponds, the basins of which may be due to collapse of the surface Quaternary conglomerate into solutional cavities in the evaporites beneath.

In Europe the largest limestone springs are those of the Stella at Castella Sacile in Frioul, northern Italy, with a mean flow of 33·6 m³/sec, but more famous are the sixteen springs of San Giovanni, 25 km north of Trieste, which merge into the big River Timavo for its short course to the sea. This is the resurgence of the Reka River of the Slovenian karst, which sinks at Skocjanska 30 km away. The mean flow of the Timavo is 26·25 m³/sec. In the United States there are the limpid but 'boiling' pools 10 m deep of Silver and Blue Springs in northern Florida, each with discharges of 14-15 m³/sec.

Karst springs generally have a more even flow than karst rivers which do not depend entirely on underground supply. The mean monthly discharges of Ras el 'Ain vary only between 0·93 and 1·08 of the mean annual flow whilst the extreme minimum recorded declines only to 0·77 of it. Flood discharges of the Silver and Blue Springs are never more than three times the mean flow. These characteristics are a product of the slowness of groundwater movement compared with surface flow, the big storages underground, and the tight bottlenecks which can restrict discharge.

However, not all of them are so well regulated. The Fountain of Vaucluse itself, which is a big spring with a mean flow of 26-27 m³/sec, has floods about seven times the mean discharge and disgorges muddy water at these times. Other large springs such as the Bourne in the Vercors plateau in the French Alps and the Vidourle near Nîmes can flood violently also. These have very open cave systems behind them with big drops. Unreliable precipitation regimes and liability to intense falls can also induce livelier springs. In a context of modest relief in comparison with the French examples, the Blue Waterholes on Cooleman Plain, N.S.W. (Fig.· 14b) have recorded a range of 0·17 m³/sec to 2·19 m³/sec and muddy water is churned out in floods. Nevertheless in virtually all circumstances karst springs damp down peak flows compared with surface channels.

Similarly springs respond to rain more slowly than surface streams but vary in this lag. A flood pulse can reach the Blue Waterholes 24-48 hours after heavy rain with the most distant streamsinks no more than 5 km away, though the peak in the

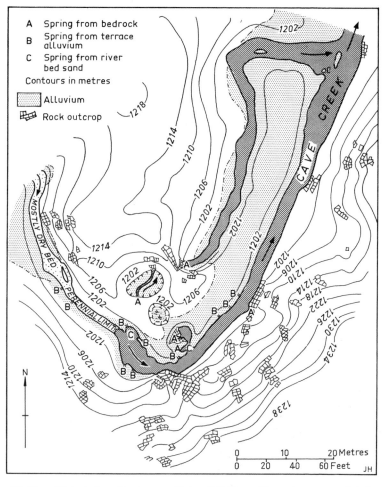

19 *Blue Waterholes, Cooleman Plain, N.S.W., with springs resurging in various manners*

surface channel they feed will already have passed. Seasonal changes in spring discharge may, however, follow the onset of a rainier season by several weeks. The caves of northwest Clare (Shaw and Tratman 1969) respond more quickly than the Blue Waterholes—in a matter of hours—probably because of freer circulation and little storage in these shallow, youthful systems. At the other extreme with its long and deep artesian flow, Ras el 'Ain changes its discharge very slowly and the control appears to be surface runoff over the intake beds one year earlier.

Karst springs have dominantly carbonate waters except where evaporites are involved as mentioned above. They vary in their response in chemical content to seasonal and weather changes as they do in their discharge regime. This is evident in the comparison Smith and his colleagues have made between risings in Mendip and in northwest Clare (Smith and Mead 1962; Smith 1965; Smith and others 1969). Three out of four major risings in Mendip have remarkably constant calcium carbonate concentrations around 235 mg/l. This is thought to be due to the tapping of substantial bodies of groundwater which has reached saturation equilibrium with high PCO_2 acquired by water percolating through grassland soil cover on Carboniferous Limestone. Inputs of streamsink water from Devonian sandstone inliers are small in comparison. The fourth rising behaves differently, varying in concentration between 120 and 200 mg/l inversely with discharge; the flood pulses carried away less limestone per unit volume. The northwest Clare risings behave in this second way with individual risings varying as much as 120-205 mg/l within a month's observations. This is attributed to a much higher proportion of streamsink water from overlying shales to seepage water from the surface of the limestone and to more direct flow from sinks to rising with a proportionately smaller phreas.

The Blue Waterholes on Cooleman Plain behave like the Clare risings seasonally, and again there is a high proportion of impervious rock to limestone in the catchment serving the risings (Fig. 14b). The Porth-yr-Ogof resurgence in South Wales (Groom and Williams 1965) is similar, varying between 31 and 75 mg/l. At low stage of river flow, the waters go underground from the Ystradfellte streamsink to the resurgence, but in flood some of the discharge proceeds beyond Ystradfellte to the entrance to Porth-yr-Ogof and has a short underground flow through that cave only. The River Mellte derives most of its water from Devonian sandstone country and this does not reach saturation equilibrium in a short course across the Carboniferous Limestone outcrop. Despite low concentrations in flood flow, the solute load is absolutely greater because of great volume. In 1960-1, 285 tons of $CaCO_3$ were removed in 38 days of flood, 274·5 tons in 175 days of normal flow, and 254·6 tons in 152 days of drought tapping groundwater only, with high concentration nearly balancing low discharge. In this way corrosion is evened out through the seasons and varying weather much more than corrasional action which is so strongly tied to floods.

Springs in the Jurassic limestones of the Cotswolds depend entirely on seepage water and Smith (1965) reports that $CaCO_3$ concentrations remain virtually constant around 300 mg/l. Since relief and climate are similar to those of Mendip nearby and the limestones of both areas are very pure, the difference in equilibrium concentrations must be sought in other lithological characters, e.g. greater porosity of Jurassic limestones or in differences in soil, perhaps with higher P_{CO_2} in the Cotswold soil air.

RELATIONS OF SURFACE AND UNDERGROUND DRAINAGE

Cave exploration, water analysis, and water tracing have permitted the relations between streamsinks and risings to be established in many cases. Sometimes these prove to be those to be expected from the surface conditions, including the disposition of intermittent streams and of dry valleys, and at other times they are not (Gèze 1958).

The simplest case is where the underground course simply takes over during normal and low flow the role of the surface course. This is the case between the Owenterbolea streamsink and St Brendans Well rising on the Poulnagallum Cave-Gowlaun River drainage in northwest Clare (Collingridge 1969). Such systems may well develop from a short underground course initially, extended by the sinking taking place at successively more upstream swallets and successively more downstream resurgences taking over from one another in the course of time, but all remaining along the line of the surface course which formerly functioned for the whole regime.

The same close relationship can persist even where the surface course ceases ever to take flood flow and the intervening dry valley has become a string of closed depressions. The waters may still emerge lower down precisely where it once flowed over the surface or very close by. Part of the South Branch of Cave Creek on Cooleman Plain has drought, normal, and flood sinks successively lower down its valley, which continues meandering onwards and deepening to join the North Branch gorge. But whereas the North Branch flows intermittently over the full length of its valley to the Blue Waterholes, part of the South Branch valley below its sinks has no channel at all though it still has a continuous downward fall. Nevertheless its waters emerge at the Blue Waterholes, some of which are actually in the bed of Cave Creek. The Bathhouse Cave creek of Fig. 32 emerges from Coppermine Cave in the Yarrangobilly valley 1·3 km away. This outflow point is about 425 m north of the junction with the main river gorge of a dis-organised dry valley which runs down from the Bathhouse Cave

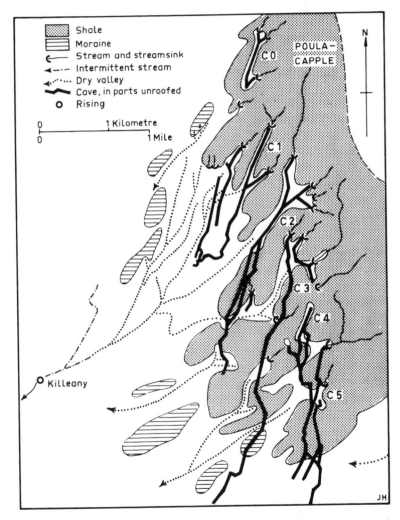

20 *Relationship of Cullaun Caves, Clare, Ireland, to surface streams and dry valleys. After Ollier and Tratman 1969.*

blind valley, but essentially there is still good correspondence between the surface and the underground drainage.

However, eventually underground flow becomes independent of the former surface flow and the structure of the limestone can very frequently be seen to be the cause of the severance. A stage of partial departure from the surface drainage pattern is exhibited by the Cullaun caves on the west of the shale ridge of Poulacapple in Clare, Ireland (Ollier and Tratman 1969). Surface streams in valleys in the shales are continued southwestwards and westwards to the Killeany valley and spring along dry valleys cut in glacial moraine down to the Carboniferous Limestone (Fig. 20). After crossing on to the limestone, streams enter shallow caves which follow the dry valleys at first. Then they are deflected SSW along major joints. In this way they pass under divides south of their originating valleys and join the Killeany waters south of the former surface drainage. Indeed the southernmost Cullaun series 4 and 5 extend beneath the shale ridge of Knockvoarheen bordering the Killeany lowland on the south. Where these waters rise is not known but it is not at Killeany.

Faults sometimes provide favourable zones for the development of underground drainage and may deflect flow along themselves till they intersect a deep valley where resurgence can take place at a point far removed from the original stream course. Gèze (1958) cites the example of the Buège, which sinks in a channel still in active use part of the year to join it to the River Hérault, but a NNE-SSW fault directs the underground flow in that direction to reappear at the Source des Cent Fons in the Hérault gorge at a point 10 km downstream of the surface junction (Fig. 21a).

In folded structures, especially where other rocks are included in them, the fold axes may deflect underground drainage away from the surface courses. Synclinal troughs in particular act in this way. At Mole Creek, Tasmania, the general trend of the drainage is northeastwards from the Great Western Tiers, the northern flank of the Central Plateau, across the axial trends of a synclinorium in Ordovician limestone and sandstone (Fig. 21b). Marakoopa Creek is one of these streams and has a surface channel across a syncline in the limestone and thence through Sensation Gorge in an anticline of underlying sandstone. However, it only rarely passes through the gorge; more frequently it sinks just short of it and follows an underground course SSE along the

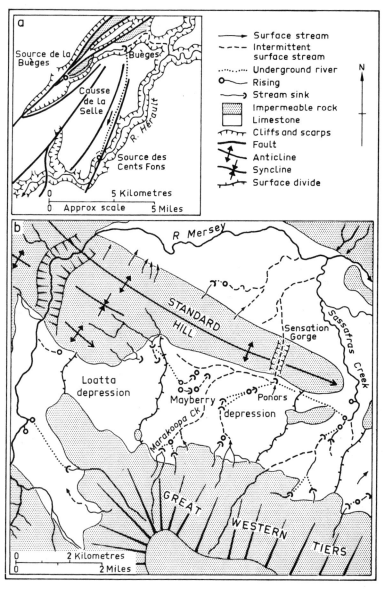

21 *Structural deflection of underground drainage from surface pattern*
(a) *By faults in the Causse de la Selle, France. After Gèze 1958.*
(b) *By folds at Mole Creek, Tasmania.*

syncline to join Sassafras Creek, which then takes the water NE
just round the nose of the pitching anticline of sandstone.

THEORY OF KARST HYDROLOGY

Although caves allow more information to be obtained about
underground water movement in karst than in other rocks, the
fact remains that the former is the less well understood. Indeed
the fundamental nature of water circulation in karst has stayed
controversial since the late nineteenth century when two conflict-
ing schools of thought developed in Europe. Since it is so basic to
the study of karst landforms, the essence of the problem will be
set out at this point, even though it must presume some of the
geomorphic data to be discussed later.

First it will be useful to outline some of the ideas which derived
from the study of permeable rocks other than those giving rise to
karst. Water, which escapes downwards from the zone of moisture
in soil, passes first through a zone of aeration where pores are
only transitorily filled with water and thence into the zone of
saturation where it displaces all the air because hydrostatic
pressure is greater than atmospheric pressure. The upper surface
of this saturated zone is the *watertable* in which hydrostatic
pressure and atmospheric pressure are equal. The watertable
parallels the land surface in subdued fashion and there is movement of the groundwater in accordance with the slope of the
watertable, i.e. down the pressure gradient. The Frenchman,
Darcy, experimented with water movement through columns of
sand to discover his law that flow through such a permeable
medium is proportional to the pressure gradient or the loss of
'head'.

$$\text{Darcy's Law} \qquad Q = K\,A\frac{H}{L}$$

where Q = flow
K = coefficient of permeability
A = cross-sectional area of flow
H = loss of head between two points
L = distance apart of these points

The total head is the sum of the head due to gravity (elevation head) and the head due to pressure of the water column.

Where a bed which transmits the water (an *aquifer*) is not overlain by impermeable materials, the groundwater is said to be unconfined and the watertable is everywhere free to rise and fall with seasonal variation in the amount of water percolating downwards.

However, the groundwater may permeate beneath an impermeable stratum (an *aquiclude*). Here the groundwater is confined and is not free to rise indefinitely with accessions of water from above. If the aquiclude is inclined and bores are put down through it, water will rise up them to heights above the confined zone. This is the artesian condition and the imaginary surface to which the bore waters would conform is known as the *piezometric surface*.

Karst watertable and single aquifer

The view of karst hydrology which until recently has dominated Anglo-American literature hinges on the concept of a watertable and is particularly associated with the name of A. Grund (1903, 1910a), who was primarily concerned with the Dinaric karst. J. Cvijić (1893) originally thought along the same lines but later modified his ideas considerably (Cvijić 1918, 1960).

It is assumed that underground water in karst behaves in essentially the same way as it does when moving through other pervious materials such as sands, gravels, and sandstones. Thus below the soil moisture zone, three relevant hydrological zones can be distinguished (Fig. 22).

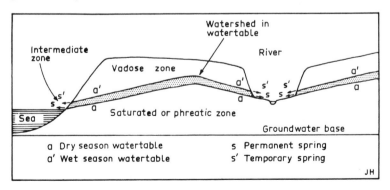

22 *Karst hydrologic zones based on concept of a watertable and a single aquifer. After Cvijić.*

(a) There is an upper or *vadose* zone in which water moves dominantly downwards after rains but which can be wholly or partly dry. It is important to distinguish between *vadose seepage,* which refers to water from rain or the soil percolating downwards, often in confined fissures, and *vadose stream flow* where water, gathered into concentrated runoff at the surface, moves streamwise downwards and laterally in open passages.

(b) In the lower saturated zone all cavities are permanently full of water and the top of the zone forms a watertable with underground watersheds similar to watertables in other rocks but thought to be flatter. This is the *phreatic zone* and this permanent body of water is termed the *phreas.* Springs occur where the water-table intersects the surface.

(c) There is an *intermediate zone* where the cavities are inter-mittently flooded to capacity, a zone through which the watertable rises and falls. This explains the behaviour of poljes (large, flat-floored closed depressions) which flood in winter when the water-table rises and which may be fed from the same openings that drain them in the dry season.

Karst conduits and multiple aquifers

The very different permeabilities assigned to different rocks (Table 1) indicate that the transfer to karst of concepts in part derived from Darcy's experiments with sand may be misleading. A critical difference is that in karst the underground water is not moving chiefly through intergranular pores (to be regarded as a single aquifer) but through narrow fissures and large caves (to be regarded as multiple aquifers: Thrailkill 1968). So it was specu-lated quite early that there is no watertable in the ordinary sense in karst. Von Katzer (1909) was one of those who maintained this for the Dinaric karst but the strongest attacks came from speleologists, notably E. Martel (1910, 1921).

Wells and bores put down in limestone close together often reach water at very different levels; dry holes occur cheek by jowl with good yielding ones. Similarly tunnels driven through limestone reveal dry and water-filled fissures close together. Frequently water-filled cavities overlie empty ones. The tracing of under-ground water movements by dyes, etc., has often shown that underground water connections can cross one another and pass under surface streams without interference (Fig. 23). In a particu-

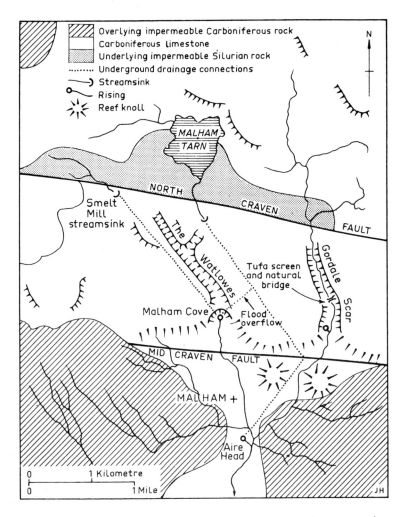

Overlying impermeable Carboniferous rock
Carboniferous limestone
Underlying impermeable Silurian rock
······· Underground drainage connections
Streamsink
Rising
Reef knoll

N

MALHAM TARN

NORTH CRAVEN FAULT

Smelt Mill streamsink

The Watlowes

Gordale Scar

Tufa screen and natural bridge

Malham Cove

Flood overflow

MID CRAVEN FAULT

MALHAM +

Aire Head

0 1 Kilometre
0 1 Mile

JH

*23 Crossing of underground and surface drainage, and other geomorphic
features at Malham Cove, Craven, England*

lar karst area, poljes at the same level behave differently, some
flooding, some not. Sometimes a high-lying polje floods before
lower ones close by. When some poljes flood, their streamsinks do
not reverse their flow but take in more water. Polje lakes have
been drained by deflecting streams from them; this could not occur
if the lakes were due to the polje intersecting the watertable. In the
particular case of Livanjsko Polje, there are three separate flat

floors, each with its own spring-surface channel-streamsink system. Artificial opening up of their streamsinks in the late nineteenth century had different effects on each, demonstrating their hydrologic independence.

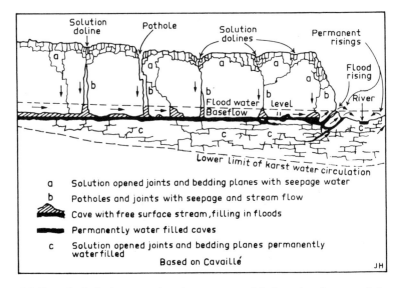

24 *Karst hydrologic system based on concept of independent karst conduits and multiple aquifers. After Cavaillé.*

Therefore Katzer, Martel and others maintained there is no watertable, only independent underground conduit systems operating like rivers but in a three-dimensional space (Fig. 24). Parts of these underground systems are free-surface streams, only partly filling caves and flowing under gravity alone. Other sections of the streams fill the caves and water can bank up behind constrictions, developing hydrostatic head. The hydrostatic pressure may drive water uphill in parts of the systems; thus in the French Alps rising water currents under pressure are known to carry pebbles up with them over heights of at least 100 m. In La Luire pothole, water can rise with a speed comparable with those experienced in surge shafts in hydroelectric systems.

Compromise views

Nevertheless, there remained difficulties for this contrary view of karst hydrology. In some areas the numbers of springs are very much fewer than points where water sinks and they lie at very similar levels despite lack of obvious structural cause for their equivalence. So elaboration of ideas about 'independent karst conduits' followed, particularly by O. Lehmann (1932) who proposed the following evolution in the hydrological system in karst (cf. also Cvijić 1918).

Initially when a dense limestone is first stripped of impervious cover or exposed to the atmosphere by uplift from beneath the sea, it will not be permeable. So to begin with a normal surface drainage with a river valley system develops over it. Then seepage water opens up planes of weakness and links up any tectonic cavities there may be; as a result the area becomes permeable. This ultimately produces a mature karst hydrology, with many independent but complicated, branching, and net-like systems of passages and chambers. Free intercommunication is not established between them and there is no single watertable. Each conduit system has many entrances but streams join underground to feed a single or a few outlets. Because of variation in cross-section, the pressure flow in filled passages is complex and not always downwards and outwards. Under hydrostatic head, water will rise higher up shafts leading from passages of large cross-section than up those from ones of small cross-section because of the inevitably varying velocities (Fig. 25). This can lead to backward flow at higher levels. Such systems may explain the incongruent behaviour of neighbouring poljes. Thrailkill (1968), however, calculates that this cross-section effect will be quite small.

Further karst development according to Lehmann involves widening of passages and removal of intervening obstacles; fissures link neighbouring caves and free-surface flow predominates in very elaborate systems. Eventually projecting limestone masses are reduced in size and riddled with passages. Lehmann regards this old stage as degenerate karst hydrology, when something approximating to a karst watertable is established. To avoid begging the question of the nature of karst hydrology, some authorities, e.g. Sweeting (1958), write of 'rest levels' instead of 'watertables', which have connotations that are inappropriate but hard to avoid.

The most important recent work bearing on this general question has been that of Zötl (1957, 1965), employing spore drift in the Austrian Calcareous Alps. He examined the underground drainage of high limestone plateaux surrounded by deep

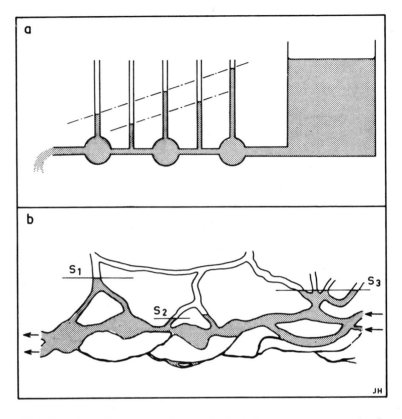

25 *Effect of varying cross-section and hydrostatic pressure on water levels. After Lehmann 1932.*

valleys, such as the Dachstein and the Totes Mountains, which cannot be regarded as in an old stage of geomorphic development. Spores fed into streamsinks near the margin of the Dachstein plateau reappeared in a few springs close together at the foot of it on one side (Fig. 26a). But when sinking waters farther and farther into the plateau were tagged in this way, springs on wider and wider sectors of the margin were reached by them. A sink in the centre of the eastern Dachstein distributed water more or less throughout it and similar broad spreads were determined for the Totes Mountains. On this evidence Zötl did not revert to Grund's simple scheme but argued that underground river systems establish interconnections earlier than Lehmann conceived. The major underground rivers remained the dominant arteries and any spores

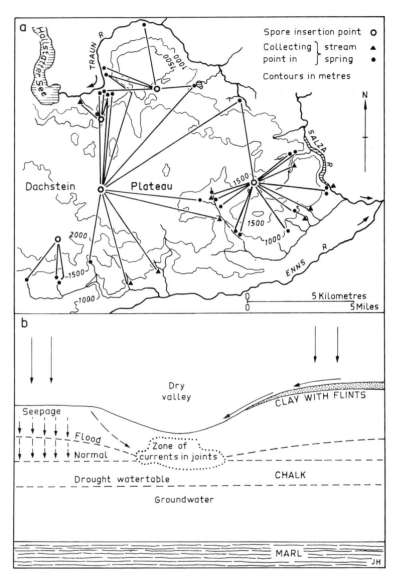

26 (a) *Underground drainage connections determined by spore drift in Dachstein plateau, Austria. After Zötl 1957.*

(b) *Underground drainage in the Chalk of Northern France. After Pinchemel 1954.*

put into streamsinks close to such arteries would reappear at a single point or a few outflow points.

Zötl's interpretation of his results resembles Pinchemel's for the hydrology of the Cretaceous Chalk country of northern France and southeast England (Fig. 26b). In this very porous limestone, evidence from bores and wells has generally been accepted as according with a single aquifer and continuous watertable concept (e.g. Balchin and Lewis 1938). However, Lewis also commented on the wide spacing of springs along the scarp foot of the Chalk south of Cambridge, England, which suggested the gathering of considerable underground drainage to feed central outflow points, possibly along major faults or joints. Pinchemel combines the two elements in his scheme, with a normal watertable reflecting the surface relief but with concentrations of linear flow along joint systems beneath the dry valleys. Ineson's maps of permeability in the Chalk of England (1962) support this concept but also show greater permeability along anticlines and domes.

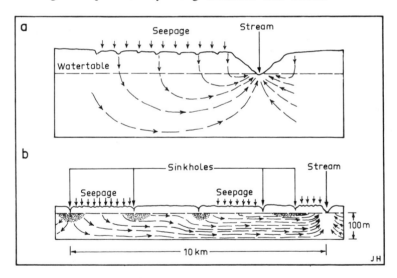

27 (a) *Pattern of deep phreatic movement. After Davis 1930.*
  (b) *Streamlines of underground flow with irregular inputs. After Thrailkill 1968.*

Deep phreatic flow

Another aspect of groundwater movement in pervious but non-karst rocks is deep flow below the watertable (Davis 1930;

Rhoades and Sinacori 1941; Fig. 27a). Deep borings in limestone beneath dam sites in the Tennessee Valley substantiated the
presence of water-filled cavities as much as 100 m below the river
bed (Moneymaker 1941). The theory is that water descends to
great depths beneath interfluves and rises back to the surface under
the valleys where springs are located. Glennie (1954a) termed
waters rising from such deep phreatic paths 'artesian'; this is
inapplicable in a strict sense but it serves to remind us that compacted limestone can virtually act as its own aquiclude. Cave
diving has shown that waters do rise 50-100 m in some major
springs but amongst European speleologists at least the view
prevails that they are fed by largely independent arteries.

A recent elaboration of the pattern of flow below the watertable
is that of Thrailkill (1968; Fig. 27b), allowing for an irregular
distribution of inputs of vadose seepage and vadose flow. This
pattern is based on the idea that important vadose flows more
distant from the valley lines will be displaced laterally and
vertically by seepage and flow inputs close to the outflows. This
concept seems to give insufficient weight to Zötl's empirical
evidence.

There is, however, a great difference in conditions between the
high alpine plateaux with complex structures of Zötl's studies and
the nearly horizontal limestone and modest relief of Kentucky
where Thrailkill derived much of his experience. Nor are these
two examples sufficient to indicate the wide range of relief,
geological, climatic and historic contexts encountered. For
example, in the Nullarbor Plain, with its horizontal limestone, low,
flat relief and a highly porous limestone at the levels at which
water rests, nearly motionless lakes in caves suggest a single
aquifer and an almost horizontal watertable very little above sea
level at distances of 20-30 miles inland in a free karst situation.
Nevertheless the occurrence of widely spaced, big cave systems
which possess these lakes again points to concentrated lines of
flow. In this instance they may be virtually inactive today and
inherited from Pleistocene periods of more effective, if not greater
precipitation. It is a case which does not fit simply into the
different schemes set out so far. White (1969) has proposed seven
kinds of karst hydrogeologic systems for regions of low to
moderate relief based on the areas he knows best in eastern and
central United States (see p. 209 and Fig. 65). It is clear that a
complete body of theory in this subject cannot be expected yet.

VI

SURFACE LANDFORMS

The larger surface landforms of karst are the subject of this chapter, in particular the special attributes which valleys take on there and the closed depressions which are the most characteristic features of all in karst. Tectonism, vulcanicity, glacial erosion and deposition, deflation and aeolian deposition, glacier and ground ice melting also fashion closed depressions but not with as much variety or as commonly. Usually these other origins are readily distinguishable but composite forms can be problematic.

Rivers traversing karst areas and those rising within them develop gorges more frequently than do those in other rocks, given similar general relief and climatic conditions. The Grandes Causses of the Massif Central of France are divided into four separate plateaux by the gorges of the Lot, Tarn, Jonte, and Dourbie, 300-500 m deep. In New South Wales, Bungonia Creek descends about 450 m in some 6 km to join the rejuvenated Shoalhaven River. In so doing, it runs initially along Devonian volcanics, including toscanite lavas, then turns at right angles to cross the line of strike, first of nearly vertical Silurian limestone and then of underlying Ordovician shales. The last give rise to a steep-sided but still V-shaped valley; a gorge is cut in the lavas but it is outshone by a slot canyon across the limestone, which has practically vertical walls about 300 m high (Pl. 15). In young folded mountain ranges, limestone gorges of much greater dimensions occur such as those of the Verdon in the southern French Alps or of the Strickland in New Guinea. However, the same habit persists in minimal local relief. In the Limestone Ranges of West Kimberley,

98

15 Bungonia Gorge, N.S.W. Canyon eroded by Bungonia Creek across strike of Silurian limestone dipping nearly vertically.

for example, there are Geikie Gorge on the Fitzroy River, Windjana Gorge on the Lennard River, and Mount Pierre Gorge on Mount Pierre Creek, amongst others; their walls are vertical but the relative relief is everywhere less than 100 m.

The prevalence of gorges in karst is primarily due to the balance between slope processes and river incision favouring the latter to an exceptional degree. Longitudinal profiles in karst tend to be flatter than in many neighbouring rocks, probably because solubility permits more lowering of talwegs than does corrasion, even in the case of massive and compacted limestones and dolomites. Dissolved load is easier to transport than clastic load and a flatter river profile results in limestone areas, and so gorges can be remarkably persistent from entry into limestone to leaving it. However, the prime factor in their formation is failure of slope processes to flare back the valley sides to a V cross-section.

Marked infiltration and reduced runoff minimise slope wash and many kinds of mass movement which would otherwise bring about such change of form. Additionally, cave roof collapse has been frequently offered as an explanation of gorges in karst terrain. This is no doubt the correct explanation in some cases— in the Rakova Kotlina northeast of Postojna, Slovenia, in which there are two natural bridges; between the arch and the main Maungawharawhara Cave in the King Country, New Zealand; between the first and second sections of Stockyard Gully Cave, southwestern Australia. However, many gorges previously regarded as collapsed caves are no longer interpreted as such. This is the case with Cheddar Gorge, Mendip, England, where surface rejuvenation forms can be traced into the gorge in which the caves are relict phreatic caves (Ford and Stanton 1968).

MEANDER CAVES

A contributing factor to the formation of gorges in karst is the effectiveness of solution in lateral action by rivers. Corrasional undercutting of valley sides encumbers a river with debris but corrosion does not. Consequently meander caves are better developed in karst than elsewhere, though it must not be thought that they are very frequent or very important landforms. A good example is provided by Verandah Cave, Borenore, New South Wales, which is situated in the undercut cliff of the concave bank of an ingrown meander of Boree Creek (Fig. 28). The Verandah itself is a remnant of a higher abandoned meander cave corresponding to a rock terrace remnant on the upstream side and is much less impressive than the 30 m deep active meander cave.

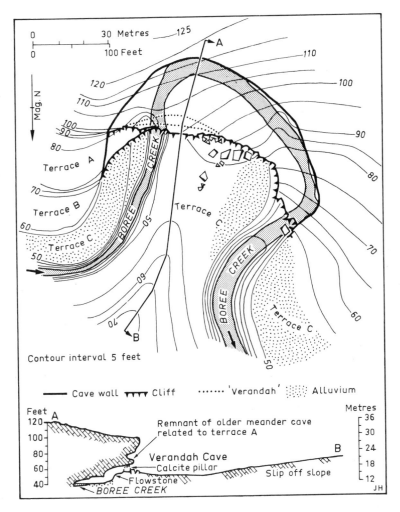

0   30 Metres
0   100 Feet

Mag. N

Contour interval 5 feet

—— Cave wall ▼▼▼▼ Cliff ·······'Verandah' Alluvium

Feet A
120
100
80
60
40

Remnant of older meander cave
related to terrace A

Verandah Cave
Calcite pillar
Flowstone
BOREE CREEK

Metres
36
30
24
18
12

B

Slip off slope

JH

28 *Meander cave on Boree Creek, Borenore, N.S.W.*

NATURAL BRIDGES

Natural bridges are more common in karst valleys than in others
but they vary much in form and genesis. Cleland (1910) made a
terminological distinction, which has not established itself in the
literature, between natural bridge through which a river runs or
has run, and natural arch where the span does not cross a valley
but perforates buttresses, spurs and ribs of rock as a consequence

of weathering, e.g. Porta di Prada, La Grigna, north Italy. The distinction between bridges and caves must be an arbitrary one. As satisfactory as any is probably the criterion that daylight reaches through a bridge. The Arch Cave at Abercrombie, N.S.W., which is 180 m long, is at about the limit of bridge on this definition. The larger and straighter the way beneath a bridge the greater length can still be lit from outside.

Some bridges are due to a very narrow band of limestone lying across a stream course. A cave developing here as the river incises is very likely to become nothing more than bridge. Steeply dipping beds are most favourable to this and the Grand Arch at Jenolan, New South Wales, is about 140 m wide in an outcrop of 180 m, dipping at 60°. The greatest span is 50 m and the maximum height is 20 m above Harry Creek which flows through the Arch.

Other bridges are the surviving elements of former cave roofs, which may have been of considerable length as in the instances of the Rakova Kotlina and the Maungawharawhara given above (Fig. 29a). Frequently natural bridges of this type seem to be associated with underground river capture with the valley below the bridge being much more deeply incised than that a little way above it. This is the explanation which Woodward (1936: cf. Wright 1936) gives of Natural Bridge on Cedar Creek near Lexington and of Natural Tunnel on Stock Creek near Clinchport, both in Virginia, U.S.A. (Fig. 29b).

Self-capture can also lead to natural bridges. The simplest case is where meander caves on one or both sides of a meander spur finally breach the wall of limestone between the two beds of the river. But generally water escaping from the river will have pierced the spur by solution along joints and bedding planes before the edge plane action of the main body of stream water will have destroyed it bodily. London Bridge, a very beautiful if small span on Burra Creek near Queanbeyan, New South Wales, provides a very straightforward example, where the creek and valley are aligned along the strike in a general way but meanders necessarily cross it. One meander spur includes a narrow band of nearly vertical limestone. The long bend was cut off by cave development across the neck of the spur where the limestone was so favourably placed (Pl. 16). An alluvial fan built by a tributary has partly filled the old meander.

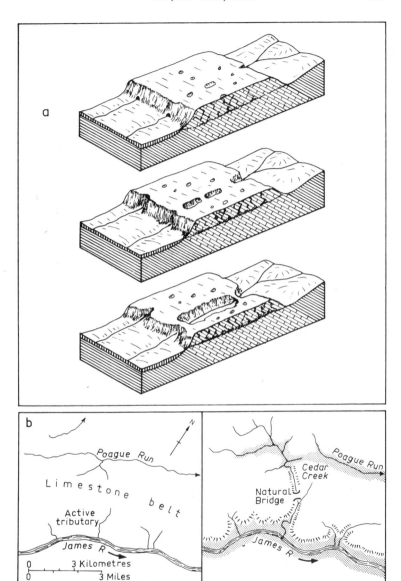

29 (a) *Natural bridges due to collapse of cave roof. After Maksimovich.*
(b) *Natural bridge due to river piracy, Natural Bridge, Cedar Creek, Virginia. After Woodward 1936.*

16 *Natural bridge due to river self-capture across meander spur. London Bridge, Burra, N.S.W.*

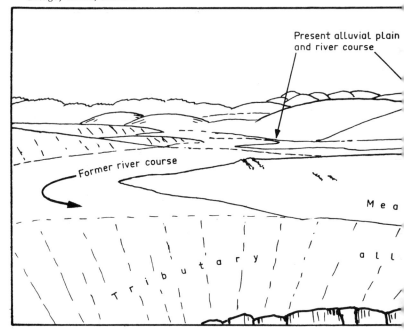

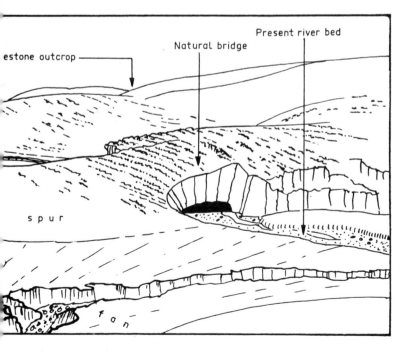

estone outcrop

Natural bridge

Present river bed

spur

fan

Bridges can also develop without meandering by self-capture at rejuvenation heads. Waterfalls can cascade down the steep drop in the longitudinal profile where joints may open up behind and engulf more and more of the flow to leave a span of rock over a retreating and degenerating fall. Cleland (1910) has proffered this as an alternative explanation for Natural Bridge, Lexington, Virginia. Other bridges are involved in karst river deposition under which heading they may be more conveniently considered.

CONSTRUCTIONAL ACTION OF RIVERS

Because of the high concentration of carbonate they may carry, karst rivers have a special capacity for constructional activity through precipitation of a portion of their chemical load. This may be due to diffusion of carbon dioxide to the atmosphere after emergence from beneath the surface, to evaporation, or to the intervention of plants. Many kinds of plants secrete calcareous skeletons or carbonate is deposited around their organic tissues and these can accumulate to form significant deposits. At Plitvice on the River Korana in Yugoslavia, Pevalek (1935) distinguishes four kinds of *tufa* formed from and around plant remains; two are dominantly formed by mosses *(Cratoneuron* and *Bryum)*, one by the blue green alga *Schizothrix*, and another depends on a grass *(Agrostis)* together with the alga. Each is associated with particular microrelief and flow conditions. In other parts of the world many other plants take on similar roles, including different orders such as the Characeae. Organic debris washed in—leaves, twigs—is also incorporated.

Tufa is the best term to use for the porous primary deposits formed in this way and *travertine* may be restricted to the more solid and crystalline carbonate deposits from flowing water, which may be secondary in nature through the infilling of the voids in tufa, or primary deposits from the water without plant participation. Another term is *calc-sinter* which some would restrict for the last category mentioned.

Plant growth, evaporation, and CO_2 diffusion are all promoted by aeration accompanying vigorous turbulence. Therefore deposition will preferentially take place where the water flows over any initial irregularity and a barrier or dam gradually builds up there (Pl. 17). This in turn favours further accumulation on the front

17 Tufa dam formation in McKinstry's Canyon, Guadalupe Range, New Mexico

where the steep slope involves frothing and bubbling, and the thinner layer passing over the actual top of the barrier is also conducive to growth. So barriers and waterfalls develop across karst rivers through their own action. These *constructive waterfalls* (Gregory 1911) may advance down the valley leaving a broader fill of travertine behind. The barriers can become overhanging with pendulous curtains of tufa and caves enveloped behind.

Alluviation may take place above the growing barriers. This has happened at the 20 m high Topolje Falls on the River Krka near Knin in Yugoslavia, above which there is an alluvial plain. The same is true lower down the Krka near Sibenik where there is a series of tufa dams totalling 40 m. However, at Plitvice the barriers have grown up too rapidly for deposition above to keep pace with them so that a number of small lakes have formed along 5 km of the valley of the River Korana (Fig. 30). The largest is nearly one km² in area and the deepest is 50 m deep. In one of the lakes

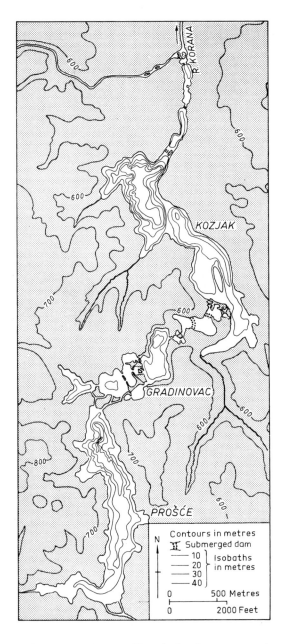

30 *Plitvice Lakes, River Korana, Yugoslavia, due to travertine and tufa dams. After Gavazzi.*

there are two submerged barriers due to a lower barrier growing higher than ones farther up valley and so incorporating two formerly higher lakes.

Phases of deposition may alternate with phases of erosion. Emig (1917) distinguishes three phases of travertine formation in the Arbuckle Mountains of Oklahoma. The first period of deposition gave rise to higher and wider falls than the present Turner Falls on Honey Creek and Prices Falls on Falls Creek.

Natural bridges occur in travertine (Cleland 1910). Some are due to actual construction of travertine above the stream through splash and spray till it meets, e.g. on Pine Creek, Arizona, but others are left in the air by the removal of underlying gravel on which the travertine was built up.

SEMIBLIND VALLEYS

Persistent sinking at a point on a river's course leads to a lowering of the bed there through bedrock solution and engulfment of sediment load. Below the sinking point the river has less power to erode its bed by solution or mechanical action. Gradually an upward step or threshold develops in the longitudinal profile and the underground course enlarges its capacity to accept more of the flow. Eventually it can take the whole flow at normal stages. Flow is continuous down the valley only after heavy rains or snow melt when the streamsink cannot accept the flood discharge. Then water banks up behind the counterslope to the level of the threshold and overflows. The bed below the threshold becomes vegetated to differing degrees and the gravels or bedrock become weathered subaerially in varying amount according to the frequency of this intermittent channel use. This is the condition of the *semiblind valley*.

Figure 31 represents an instance from Cooleman Plain, New South Wales, where a small creek flowing from a granodiorite hill sinks in normal conditions after flowing for 150 m over the outcrop of Silurian limestone into an earth hole in a small, circular alluvial flat. A low, grassy threshold, about 3 m high, separates this depression from a gravel stream bed which is only rarely followed by overflow.

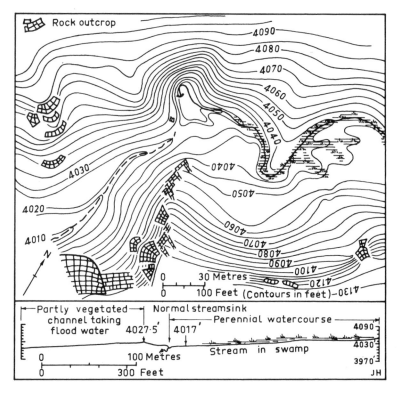

31 Semiblind valley on Cooleman Plain, N.S.W.

BLIND VALLEYS

Eventually a sinking stream cuts down its bed so far and enlarges its underground course to such an extent that the stream is always completely engulfed and never flows beyond. Thus a blind valley is produced, closed off at its downstream end. The height of the closure may be only a few metres in the case of a small stream but may reach into the hundreds of metres, particularly with large streams. The larger the stream the more likely it is to disappear into a penetrable cave. After floods, a temporary pond or small lake may form in a blind valley. The example of Fig. 32 is one of many at Yarrangobilly, New South Wales, where a strike belt of Silurian limestone forms a plateau between the Yarrangobilly River gorge and the steep slopes of a porphyry range. This stream sinks into the Bathhouse Cave below crags in a steep counterslope

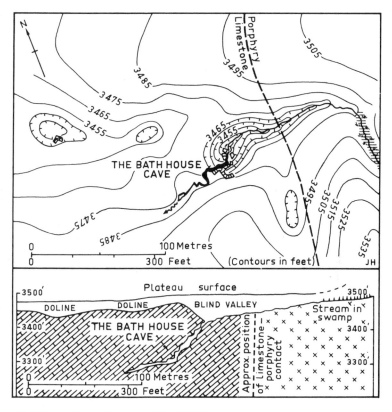

32 Blind valley at Yarrangobilly, N.S.W.

15 m high. The blind valley closes off only about 45 m from the limestone and porphyry contact. It is continued by a shallow valley some 8 m deep in the limestone plateau and interrupted by small closed depressions. This is the much modified former onward course of the stream. Sometimes a blind valley has a series of closed depressions into which the stream spills successively after banking up in flood, each one providing additional entry into the underground conduit. There are several cases of this on Cooleman Plain, N.S.W.

Blind valleys sometimes reach far into the karst area and are cut deeply below its surface. Many factors such as the characteristics of the karst rocks, the discharge and the chemistry of the sinking stream control the dimensions of blind valleys. Gams (1962, 1965) has made a comparative study of a number of blind valleys in the Slovenian karst, some of which are formed by streams flowing from headwaters on the impervious Eocene Flysch. Some of these reach depths of 200 m. He concludes that there is a relationship between the total carbonate content of the stream where it crosses on to the limestone and the nature of its blind valley. Those with low carbonate contents have longer and wider blind valleys; the ratio of valley lengths before and after crossing on to the limestone is 3·7:1. In a second group with much higher carbonate concentrations, the blind valleys are narrower and shorter with an equivalent ratio of 11:1. The width of blind valleys may also vary according to their alluvium. Where a blind valley acquires an impervious alluvial fill, this will seal off the bottom from corrosion which then becomes directed against the sides of the valley, thus widening it. With pervious alluvium such as much of the Pleistocene cold period gravels of the Yugo-slavian karst, solution can still go on beneath the openwork fill, continuing to deepen the valley without much widening.

STEEPHEADS

Some minor geomorphic aspects of springs have already been touched upon in Chapter V. Perhaps their most important mor-phological trait is the way springs, both of exsurgence and resurg-ence type, occur at the heads of valleys which begin very abruptly (Pl. 18). These are usually short valleys in the margins of plateaux or on the flanks of mountain ranges. The Fountain of Vaucluse in Provence arises beneath a cliff of 200 m height at the head of such a valley as is implied in the derivation of the name—*val clos*. Another popular name for such cul-de-sac valleys in several languages is 'World's End' but the American term *steephead* is becoming the customary word in the English language. Often these valleys are gorge-like for some distance downstream and frequently impermeable underlying rocks crop out along the bottom beneath cliffs of limestone or dolomite. A distinction is sometimes made between steepheads incised to an impervious basement and pocket valleys of the same general nature but within the karst rock outcrop.

Steepheads can form in more than one way. The form may be due solely to spring sapping. A spring undermines the slope or cliff above it, rock and soil gravitate into the spring and are removed in solution or as clastic load. In this way there is headward recession of the valley. Alternatively the gorge may be the result of collapse of substantial lengths of cave roof. In this case it may not have developed headward but in an irregular manner,

18 Spring at Sogöksu, Turkey, rising at head of short, blunt valley or steephead

19 Dry valley, Cooleman Plain, N.S.W., sunk below corrosion plain of Silurian limestone. Valley contains Pleistocene periglacial-fluvial deposits and small dolines occur along valleysides.

separate sections of collapse being eventually united. If collapse is recent, evidence in the form of natural bridges may survive but if it is ancient, all such evidence may have disappeared and criteria to distinguish between the two modes of origin may be lacking.

DRY VALLEYS

Dry valleys are similar in many respects to ordinary river valleys but there are no stream channels in their floors (Pl. 19). They are by no means restricted to karst; on many rocks there are short dry sections at the heads of valleys subject to throughflow and minor tributaries may be entirely of this nature, channels only appearing when overland flow gathers sufficiently to cut and maintain them. Dry valleys can be longer and may form branching systems on other permeable rocks such as sandstone and pumice, but these characteristics are accentuated on limestone and dolomite.

One kind of dry valley usually presents little difficulty of inter-

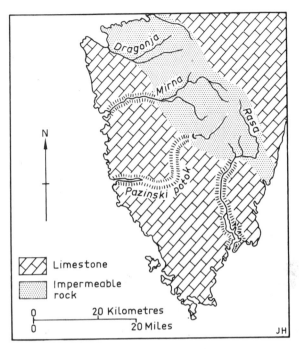

33 Dry valley of Pazinski Potok, Istria, Yugoslavia. After Roglić 1964a.

pretation but can be the most spectacular of this category of landform. After sinking on entry into karst terrain, allogenic river valleys are often continued by dry valleys, which may pass right through the pervious rock outcrop. In the Istrian plateau (Roglić 1964a), three river systems, which have eroded wide valleys in the Eocene Flysch, drain southwestwards into the limestone (Fig. 33). The northern, the Mirna, and the southern, the Rasa, cross the karst in canyons to the sea. The middle system sinks at the limestone but Pazinski Potok, a dry valley of canyon form, carries on to the coast.

Very many karst regions can provide similar examples. In Craven, the Watlowes is a craggy dry valley whereby the stream from Malham Tarn on Silurian slates formerly flowed across the limestone to cascade over the 75 m cliff of Malham Cove (Fig. 23). In south central Java, the Sadeng dry valley winds through the Gunung Sewu conekarst to the south coast (Fig. 34). It is the former course of the Solo River now flowing out of the Batoeretno basin north of the karst (Lehmann 1936). Beginning 20-25 m

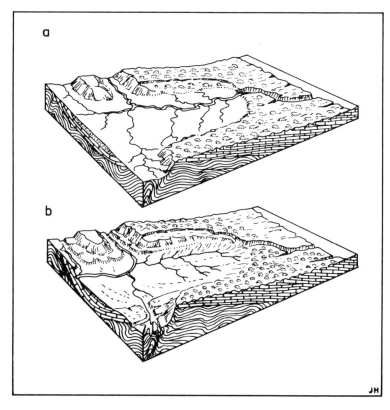

34 *Reversal of River Solo in Batoeretno basin, Southern Java, causing Sadeng dry valley through the Gunung Sewu conekarst (a) before reversal, (b) after reversal. After Lehmann 1936.*

above this basin, the dry valley is 250-300 m wide and 150 m deep; its long profile's southward fall is interrupted by elongated basins with shallow lakes or *telagas* separated by gentle swells as much as 30 m high. Figure 32 shows the beginning of a dry valley system which continues the line of the Bathhouse Cave blind valley to the Yarrangobilly River gorge. This is much interrupted by transverse rock barriers and shallow closed depressions, some of which represent former sinking points of the creek when it flowed further across the limestone plateau.

Semiblind valleys may be continued by dry valleys instead of by intermittently active stream beds when overflow is infrequent enough to maintain a channel. Water may rise up through the bottoms of depressions along such dry valleys when extreme flood

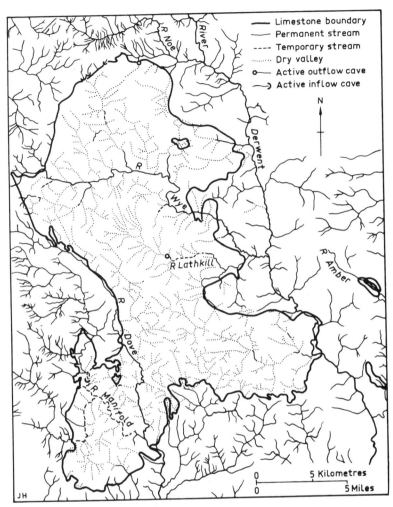

35 *Dry valley systems, Peak District, England. After Warwick 1964.*

conditions re-establish an onward flow. This is the situation with
the Owenterbolea streamsink-St Brendans' Well rising system
mentioned earlier (p. 84). Upper St Brendans is the chief intermittent rising along the dry valley.

More problematic are the branching systems of dry valleys
within karsts where there is an absence of stream channels
throughout. Most of the Peak District is covered by such systems
and there are few river channels in the area other than the

allogenic Wye, Dove, Manifold, and Derwent (Fig. 35). These dry valleys often begin in shallow, bowl-like basins, which develop into rock-walled valleys and gorges. Many smaller dry valleys hang above major dry valleys or the allogenic river valleys. The general pattern of the dry valleys resembles that of surface streams on surrounding impervious rocks, and nickpoints on dry valley tributaries of the Manifold can be related to phases of rejuvenation witnessed by the landforms of the major valley. For these reasons Warwick (1964) ascribes the dry valleys to inheritance from a former cover of impervious shales. After this cover was stripped, the rivers incised into the limestone beneath until solution opened up planes of weakness in it to permit drainage to go underground. The hanging condition is a product of main valleys continuing to cut down after tributaries ceased to have a surface flow. Hanging valleys are frequent landforms in karst; Sweeting (1950) describes them from Craven. On Cooleman Plain several dry valleys hang above the North and South Branches of Cave Creek. Discordance of tributary junction occurs along the dry valley as well as the perennial and intermittent flow sections of the South Branch.

The dry valleys of the Cretaceous Chalk of eastern England and northern France have occasioned more attention than any others. Valley systems dominate the relief of the cuestas and low plateaux in this weak and porous limestone, closed depressions are scarce and small. Allogenic rivers pass through, often in consequent water gaps, but otherwise only the lower parts of more important valleys have stream channels, usually over alluvial flood plains. Very rounded cross-sections are characteristic of the dry valleys, though thousands of years of cultivation have tended to flatten the floors and cause breaks of slope along the valley sides (Sparks and Lewis 1957). They drain both dipslopes and scarps and sometimes exhibit rectangular patterns suggestive of joint control. Exceptionally wet seasons can cause surface flow over their grassy floors.

It is difficult to attribute the Chalk dry valleys as a whole to inheritance of surface drainage from overlying impervious rocks though some have this nature near the margin of Tertiary covers. Some of the English Chalk country was planed off by an Early Pleistocene marine transgression across which dry valleys are as well developed as elsewhere. Some writers have attributed them to glacial meltwater streams, mistaking the anthropogenic flat floors for the characteristic trough cross-section of meltwater overflow channels. Though there are a few dry valleys of this origin in the Chalk of England, it is impossible as a general explanation because much of the Chalk was unaffected by glaciers and elsewhere the dendritic patterns of dry valleys do not fit any conceivable pattern of ice retreat.

Periglacial conditions have been called on, with permafrost inducing surface flow of snow meltwater and summer rain to cut the valleys (Reid 1887). Dry valleys commonly have deposits of coombe rock—angular chalk rubble in mud matrix—which are at least in part periglacial solifluction phenomena and the rounded chalk forms are attributed to periglacial masswasting as a whole by many investigators. There seems little doubt therefore but that Pleistocene phases of frozen ground and surface flow have affected very many Chalk dry valleys (Brown 1969). However, it seems unrealistic to assume plane surfaces over the whole Chalk terrain prior to the Pleistocene cold periods.

Other authorities have sought their origins in changes in the watertable. Sparks and Lewis (1957) made a strong case for interpreting some at least of the scarp dry valleys, which are steeper sided and blunter headed than the dipslope valleys, by spring sapping, followed by lowering of the watertable and down valley migration of springheads, now countersunk in the dry valley floors well below their heads. The Pegsdon dry valleys in Hertfordshire have valley fill yielding molluscs indicative of a Postglacial warm period during which springs emerged higher up the valley. However, they inclined to think that much of the spring action creating the valleys belonged to earlier periods than that, perhaps of more effective precipitation than at present. Lack of statistical correlation between joint and dry valley directions has been used by Brown (1969) to argue against spring sapping as the mechanism of dry valley formation in the scarps.

Chandler (1909) and Fagg (1923, 1954) explained the dip-slope dry valleys of the Downs of southeast England by scarp retreat which lowered the springline at the scarp foot at the contact of the chalk with underlying clay and thus also the water-table, causing surface drainage to go underground on the dipslope. But this hypothesis implies that the scarp dry valleys must be of a different later generation; for this there is no evidence. More-over, erosion surface remnants below the North Downs scarp show it has not retreated much for a long time. However, Sparks (1961) points out that suspended erosion surfaces are evidence of rejuvenation in the clay vales which must have lowered watertables in the cuesta of chalk without scarp retreat and so could have dried out both scarp and dipslope valleys.

The long profiles of dry valleys in the Chilterns show alternating graded reaches and nickpoints due to successive rejuvenations. C. D. Ollier and A. J. Thomasson (pers. comm.) consider that these facts deny genesis by meltwater runoff over frozen ground which should result in a simpler profile, but support the idea that headward sapping consequent on a fall of base level dried out higher reaches of the valleys. Similarly, Pinchemel (1954) relies on valley deepening over a long period of time to dry out tributary valleys. Springs would shift down subsidiary valleys as each incision in major valleys lowered the watertable. He terms this general process one of auto-desiccation. It can be measured by his index of desiccation, which is the ratio of valley density to stream channel density (cf. Williams (1966b) dry valley/stream valley ratio). Chalk country near Amiens gives a high ratio of 7·4 whereas Limousin granite country had an index of 1·3 and Jurassic clays, marls, limestones and sandstones of the Bas Boulonnais one of 1·85.

DOLINES AND COCKPITS

The simpler forms of karst closed depression are now commonly considered under the name 'doline' of Serbo-Croat origin; English words such as sinkhole, swallet, and swallow hole having very loose connotations. Dolines are generally circular or oval in plan, with depths varying very much in relation to diameters. There is thus a range of forms — dish and bowl shaped, conical and cylindrical. When in bedrock they approach the form of a shaft, there is transition to the pothole, which is best considered along with caves. When they become elongated in plan, there is again transition to the karst corridor (Pl. 37). If a stream runs into them, the gradation is to the blind valley.

In size they vary from a few metres in their dimensions to more than a hundred metres in depth and to several hundred metres in horizontal dimensions. With increasing size, there is usually complexity of form, which takes them into other categories of closed depression.

Bedrock crops out on the floors and sides of some dolines, which are often then more angular in form, but others are surfaced largely or entirely by soil or superficial deposits. Still others are formed in bedrock formations overlying karst rock. If there is a flat floor, this consists of detrital materials, often impervious, including insoluble clay residues. Swamps and intermittent or permanent ponds may occur on these impervious seals. In other dolines there are cave entrances or open shafts.

Several processes are responsible for doline formation: surface solution, cave collapse, piping, subsidence and stream removal of superficial covers. These often occur in combination. Nevertheless, it is useful in first analysis to consider dolines in a five-fold categorisation on the basis of certain dominant mechanisms (Cramer 1941).

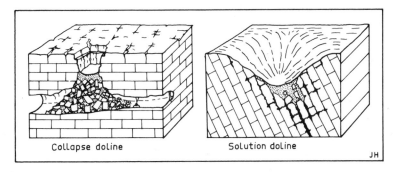

Collapse doline   Solution doline   JH

36 Block diagrams of collapse and solution dolines

(a) *Solution dolines* (Fig. 36). These are due primarily to pronounced surface solution of the karst bedrock around some favourable point such as a joint intersection. The solutes and some insoluble residues are removed down solution-widened planes of weakness, though once the latter are enlarged to shaft dimensions there will be sliding and falling of residues and rock fragments brought to their apertures. As soon as a focus of downward percolation is established by solution, it will gather drainage to itself and the embryonic doline will further its own development. In fairly uniform rock the interaction of solution, mechanical slope wash and mass movements of material with the angle of the doline sides can result in conditions of dynamic equilibrium on uniform slopes. A conical shape is therefore characteristic of dolines of this kind. However, residues may accumulate at the bottom too rapidly for removal down widened joint planes, so that they form flat and perhaps swampy floors or even cause pools. The slopes around the floor may maintain a constant and characteristic angle, usually in the 30-40° range. But if water is deflected laterally and attacks the side slopes as the bottom rises through the accumulation of impervious clays, the bedrock walls will lose their uniform angle (Aubert 1969). The character of a doline will thus depend on the two ratios:

$$\frac{\text{vertical solution}}{\text{lateral erosion}} \times \frac{\text{evacuation of solutes and clasts}}{\text{clastic fill}}$$

The origin of a solution doline will be apparent if bedrock is exposed over much or all of its surface (Pl. 20), as commonly happens in high mountains or dry country karst or where forest clearance has been followed by loss of soil and litter from the doline slopes. If soil or waste mantle covers the sides and bottom,

20 *Solution doline with bedrock exposed over floor and sides of closed depression. Velebit Mountains, Yugoslavia.*

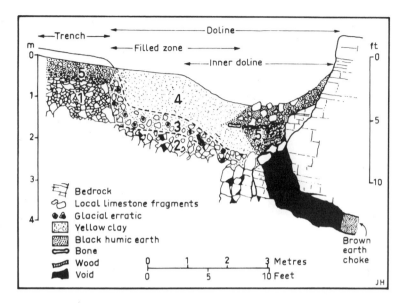

37 Section through a solution doline at Grandes-Chaumilles, Swiss Jura. After Aubert 1966. Numbers explained in text.

origin by surface solution may still be witnessed by a shaft or solution pipe in undisturbed bedrock in the bottom of the depression.

Aubert (1966) has sectioned a solution doline at Grandes-Chaumilles in the Jura, revealing a bedrock solution pipe down which the doline contents descend (Fig. 37). Frost shattered rock (1) and residual clay soil with local rock fragments (2) underlie clay soils containing Würm glacial erratics as well (3) and stone-free clays above (4). This fill was partly evacuated in Postglacial times and the inner doline thus formed has been partly refilled by rendzinas containing talus from the doline wall, bones and wood (5). Thus the form and contents of the doline register its history from the pre-Würm interglacial or a Würm interstadial.

Symmetrical form results from uniform slope processes on uniform rock but asymmetry can arise in a number of ways (Fig. 39d). If the bedrock has a substantial dip, the dipslope side of the doline is likely to be less steep than the antidip side, bedding planes having a stronger influence on the one and joint planes on the other. Again if the dolines develop in a steep slope, the upslope sides are likely to gather more water than the downslope sides and to be reduced in angle by greater solution. In conditions of prevalent snow drifting, the leeward sides of dolines will gather more snow which will persist longer. These may thus get degraded more than the snow-free flanks. However, the opposite may be the

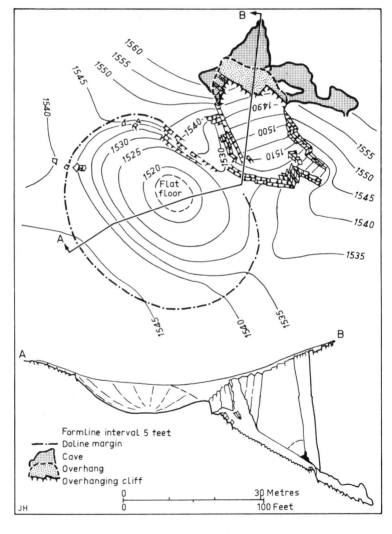

38 *The Punchbowl, Wee Jasper, N.S.W. Collapse doline on right; closed depression on left may be a solution doline.*

case if the snow-free sides have much more humus on them than the snowbank side, solution beneath the organic cover being greater than beneath the snow.

(b) *Collapse dolines* (Fig. 36). The prime cause of a doline may be collapse of a roof of a cave formed by underground solution. To begin with, these will be mainly vertically walled and often angular in plan through the influence of joints (Fig. 38). Moreover, the depth-width ratio may frequently be greater than with solution dolines where it is not likely to exceed 1:3·5.

However, unless there is further collapse, these dolines will change progressively to a conical or bowl shape through the wearing down of the sides and filling of the bottom (by solution, rockfall, frost wedging and salt wedging according to climate), soil creep once soil forms, and other processes. In time overt signs of collapse are lost and a superficial similarity to other dolines prevails. Only examination of the cave below or excavation in the doline bottom can reveal the origin.

In a dry climate, the original form may persist a long time. In the Nullarbor Plain, for example, all the large dolines seem to be of collapse origin (Jennings 1967c; Pl. 31). Where collapse has been into water-filled caves or there has been subsequent rise of water level, the collapse doline may have a lake, which can be deep, occupying its floor (Fig 40a). Such are the cenotes of Yucatan, the 'obruk' lakes of the Turkish plateau (Pl. 21), and similar features in the southeast of South Australia, including Piccaninny Blue Lake and Goulden's Hole. Divers have descended 55 m into the latter.

(c) *Subjacent karst collapse doline* (Fig. 39b). Cave collapse occurs in karst rocks beneath overlying bedrock formations. Initially steep-walled, deep dolines may result but weathering will turn them into conical features which may be degraded into still gentler forms in due course. Thomas (1954) has shown that in South Wales there are more and larger dolines on the Carboniferous Millstone Grit, a conglomeratic sandstone, than on the outcrops of the Carboniferous Limestone that extends beneath it. He attributes them to collapse into caverns in the subjacent limestone; however, there is the difficulty that bigger caves than any known in Britain as yet must be inferred from the size of the Gritstone depressions.

21 Collapse doline with 'obruk' lake near Konya, Turkey. A cenote.

No other explanation seems possible for the Big Hole near Braidwood in New South Wales where Silurian limestone is inferred to underlie a Devonian quartz. sandstone in which this 115 m hole with a high depth/width ratio is found (Fig. 40b). Here a chamber large enough to account for a cavity the size of the Big Hole is known in Wyanbene Cave nearby beneath the Devonian-Silurian unconformity, with shafts stoping to the overlying sandstone (Jennings 1967b).

(d) *Subsidence doline* (Fig. 39a). Where superficial deposits or thick residual soils overlie karst rocks, dolines can develop through spasmodic subsidence and more continuous piping of these materials into widened joints and solution pipes in the bedrock beneath. They vary very much in size and shape. A quick movement of subsidence may temporarily produce a cylindrical hole (Pl. 40) which rapidly weathers into a gentler conical or bowl-shaped depression. More or less continuous small-scale movements may initiate and maintain forms of this second kind.

The 'shakeholes' of Craven are conical subsidence dolines in the glacial moraine left on the limestone by the final Pleistocene

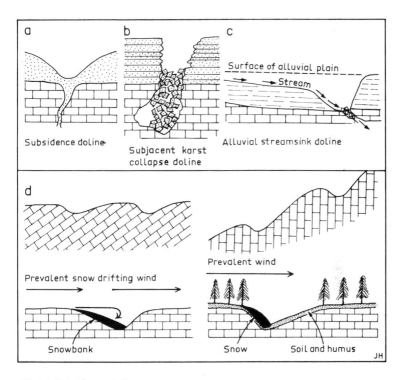

39 *(a) Subsidence doline.*
   (b) Subjacent karst collapse doline.
   (c) Alluvial streamsink doline.
   (d) Some causes of doline asymmetry.

glaciation (Sweeting 1950). In the Mole Creek area of Tasmania, many dolines occur in Pleistocene gravel fans, of glacifluvial and periglacial-fluvial nature, resting on Ordovician limestone (Jennings 1967a). Lowry (1967a) has shown that the conical doline feeding into open solution pipes above Easter Cave, Augusta, southwestern Australia, is a subsidence doline in thick residual soils above aeolian calcarenite (Fig. 7b). Dolines can also be formed by solution at the top of subjacent karst rocks and by gradual subsidence or recurrent spasms of subsidence of overlying weak bedrock through progressive removal of their support. Cramer (1941) and Hundt (1950) cite many areas in central Germany where this happens above evaporites.

(e) *Alluvial streamsink dolines* (Fig. 39c). Dolines form in alluvium where streams sink into underlying karst rock. The

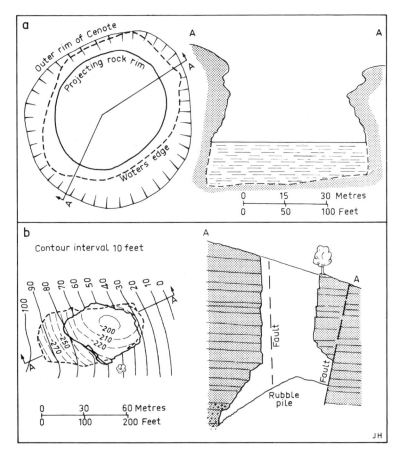

40 (a) *Hells Hole, Mt Gambier, South Australia, a cenote. After Cave Expl. Grp S. Australia survey.*
(b) *Braidwood Big Hole, N.S.W., a subjacent karst collapse doline.*

processes which create subsidence dolines operate here but additionally the stream provides a good channel for mechanical removal of the insoluble alluvium. Stream-cut trenches lead into the side of otherwise conical dolines produced in this way. This kind of doline is frequently blocked by detritus and much of the time there may be no karst bedrock visible at all.

Because dolines often occur in large numbers close together (Fig. 41, Pl. 22), geomorphologists were early led to rudimentary quantitative analysis in their study (e.g. Lozinski 1907). Drawing on these earlier efforts and adding to them, Cramer (1941) was

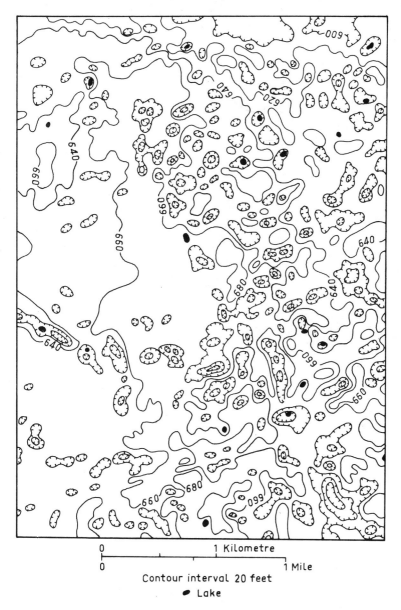

0 ⊢ 1 Kilometre

0 ⊢ 1 Mile

Contour interval 20 feet

● Lake

41 Doline karst near Mammoth Cave, Kentucky. Drawn from U.S. Geol. Surv. map.

22 *Doline field at Craigmore, South Canterbury, New Zealand. Dolines along joints and along dry valleys, partly due to solution in Oligocene limestone and partly due to subsidence in loess cover. Photo by E. Thornley by courtesy of New Zealand Geological Survey.*

the first to venture more consciously in this direction in a cartographic analysis of doline fields in twenty areas. His measures were: mean size in the doline population, the doline density (numbers per km^2), and total doline area per km^2. Williams (1969) proposes an index of pitting which is the reciprocal of this last measure.

Cramer determined mean sizes ranging widely between 17 m^2 and 159,200 m^2, densities of $0.57/km^2$ to $2460/km^2$, and total doline areas of 0.06 m^2/km^2 to 299,000 m^2/km^2. The sample areas, selected on the basis of available detailed maps, varied lithologically (limestone, dolomite, gypsum), climatically (alpine, cool temperate continental, humid subtropical, tropical semiarid, and tropical humid), and in other ways. The dolines themselves varied genetically, though solution and subsidence dolines were regarded as the dominant types except in the gypsum karst where subjacent karst collapse as well as subsidence dolines were prevalent. With data complex enough to invite factorial analysis today, Cramer did not argue very much from his results. He recognised that there was climatic as well as geological influence on the mean size and density of both subsidence and solution dolines without being able to disentangle them. The doline fields consisted of (a) few, small dolines; (b) many, small dolines; and (c) few, large dolines. However, these types do not represent an evolutionary sequence, for which relief energy would be a better indicator. Size and density are instead measures of corrosion intensity and total doline area per km^2, the product of these two measures, is a single index for this. His results show this is highest in humid and semiarid tropical and subtropical instances and next greatest in alpine karsts. The lowest values came from cool and warm temperate continental interior examples. Many more such analyses would be necessary to substantiate these relationships.

Relief energy can be measured by depth/diameter ratio and this was early used by Cvijić (1893) descriptively. Considering small depressions of part of the Mendip plateau, Coleman and Balchin (1959) argued that if they were of solutional origin, there should be a tendency to dynamic equilibrium in their slopes and therefore a depth/diameter graph should cluster them along a straight line. Collapse dolines on the other hand would be initially variable in this ratio and weathering back of their sides would increase this variability; a depth/diameter graph would have a wide scatter as a result. A plot of 140 measured Mendip depressions gave such a scatter. It was argued that this pointed to a collapse origin but the initial data included many old mining holes. Subsequently Ford (1964) examined a larger sample of 566 definitely natural dolines. Eighty per cent of them lay in dry valleys where their longitudinal profiles had less than a certain critical angle. Thus the dolines formed by surface solution where most surface water was concentrated but where it did not run off too fast to percolate underground. Mapping of the caves in the area showed that they do not underlie the dry valleys and collapses in the caves do not correspond with surface depressions.

Measurement of a field of ninety-four dolines on the Craigmore plateau, South Canterbury, New Zealand (Pl. 22; Fig. 42), showed that there is a strong modality in their depth (mean 5·9 m, standard deviation 2·3 m). Also there is a strong correlation between depth

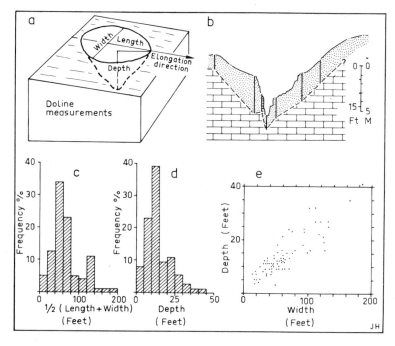

42 *Dolines at Craigmore, South Canterbury, New Zealand*
   (a) *Dimension measurements.*
   (b) *Augered profile across one doline.*
   (c) *Frequency distribution of mean diameter.*
   (d) *Frequency distribution of depth.*
   (e) *Dispersion diagram of depth against width.*

and mean diameter ($= 0.84$, significant at 0.1 per cent confidence level). These shape characteristics argue against a collapse origin and point to parallel slope retreat in them, which is compatible with solutional or subsidence origin. Boring across one of these dolines indicated a composite origin, with a younger subsidence doline in loess covering a solution doline beneath.

These New Zealand dolines tend strongly to circularity (mean length/width ratio or elongation ratio (Williams 1969; La Valle 1967) 1.32, standard deviation 0.32) and there is little tendency to a preferred orientation in the 25 per cent sufficiently elongated for the direction of elongation to be measurable. In south central Kentucky over several limestones, La Valle (1967) found there is significant elongation. He calculated mean elongation ratio for

dolines in sample areas and also the percentage of dolines elongated along joints or faults. Multiple regression analysis was used to relate these doline characteristics to other karst parameters. The two traits were found to be strongly associated with one another; the more elongated they are the more frequently aligned along structural lineaments. Both elongation ratio and structural alignment increase with purity of the limestone. They are also directly related to the gradient of the underground drainage as calculated by the karst relief ratio $\dfrac{\text{greatest height above lowest rising}}{\text{distance from underground watershed}}$.
This is thought to be due to more rapid and more efficient removal underground of solutes and residues from the limestone where gradients are greater. Also the two characteristics are more pronounced the nearer the sample area is to a rising; underground drainage will be better developed near the outflow points and so lead to more pronounced adjustment to structure in the closed depressions above. In addition percentage of structurally aligned depressions, and probably also mean elongation ratio, is correlated with the percentage of each sample area in closed depressions (cf. total doline area/km^2); the greater the development of dolines the closer their adjustment to structure in length and direction.

In tropical humid karst there are many simple closed depressions which differ from those already described in important ways (Fig. 43). They are star-shaped, with their sides lobed convexly inwards and with gullies between which carry streams after heavy rain, as Lehmann (1936) described from south-central Java. Moreover, they do not perforate a fairly simple surface but are set amongst steep residual hills. The Jamaican name 'cockpit' has been associated with this kind of closed depression. The slopes of these residual hills drain directly into the cockpits and the whole area belongs to the closed depressions. Aub (in press b) has shown that 60 per cent of the cockpits of a Jamaican karst area have bedrock shafts at or near their lowest points. This is probably true also of most of the remainder so that here there is a field of closed depressions due to surface solution.

For their morphometry, Williams (1969) draws their boundaries over the tops of the enclosing hills and through the cols between these so that all the slopes draining into the cockpits are included. From air photographs he analysed samples of various kinds of karst in eastern New Guinea using such limits. Such

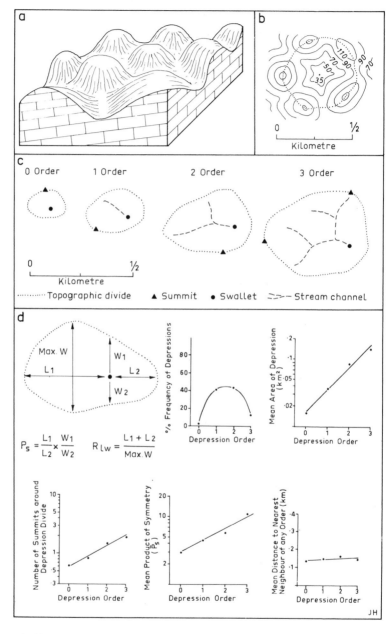

43 *Characteristics of tropical cockpits, ordering, dimensional analysis and some resulting relationships from karst on Mt Kaijende, Central Highlands, Australian New Guinea. After P. W. Williams.*

depressions were ordered on the basis of their gully systems as if they were normal drainage basins and various dimensions were measured, including distances from the lowest point in each depression to the periphery along the major axis and along one at right angles. Analysis of these dimensions revealed system in a kind of relief previously described as chaotic. For example, in karst on Mt Kaijende in the Central Highlands, the frequency of depressions was greater in intermediate orders than in the simplest and the most complex orders. Their mean area, however, increased steadily with order and the number of residual hills on their boundaries also. There was not much change in elongation $\frac{\text{length}}{\text{max. width}}$ with order nor of distance from nearest neighbour in the same order. On the other hand the product of symmetry, a measure of departure of the lowest point from a central position, increased substantially with order. Analyses of these kinds impose constraints on hypothesising about their origin and evolution.

Matschinski (1962) has devised a tensor analysis to determine any alignments in a pattern of dolines but it cannot be applied to close-set fields of dolines for which more complex tests must be devised.

UVALAS

This Serbo-Croat name has also passed into international usage to refer to complex closed depressions with more than one hollow in their make-up. Size is not a criterion but perforce they are larger than small dolines because they are formed by the merging of the simpler type of closed depression. When the rocks are dipping substantially, uvalas are generally elongated along the strike, with a chain of dolines in this trend; Cvijić (1960) cites Ceteniste uvala in Triassic limestone in southwest Serbia. Faultlines can also occasion uvalas; the depression in which Haggas Hole is situated in the upper Waitomo valley, New Zealand, is of this nature. In horizontal beds, uvalas are likely to be more lobate in plan than those structurally aligned ones.

Cvijić (1960) attributes uvalas to surface solution, but little work has been done on their formation and other modes of origin cannot be excluded. Northeast of Mole Creek, Tasmania, there is a uvala 250-350 m in diameter and 25 m in depth, made up of

14 hollows of varying size (Jennings 1967c). It forms an inlier of Ordovician limestone in a Tertiary basalt cap of the divide between the Mersey River and the catchments of two tributaries, Mole Creek and Lobster Rivulet. There has been relief inversion. A small stream drains into the deepest hollow and gullies its flank. Here it is possible that cavernous development beneath the basalts played a part in forming the depression.

Complex depressions of uvala type occur over the course of underground rivers as in the case of River Mangapu in the King Country, New Zealand, in the neighbourhood of The Lost World, a large collapse entrance into the river cave below (Pl. 23).

23 Uvala in Oligocene limestone at The Lost World, King Country, New Zealand

POLJES

In the Yugoslavian karst there are many large closed depressions with flat floors across which streams flow (Fig. 44). They are called poljes and this term is now used generically for such features, though there are many local names for them, e.g. 'plans' in Provence, 'wangs' in Malaya, and 'hojos' in Cuba (Lehmann and others 1956). They are usually elongated along the strike and tectonic axes (Pl. 25), but can also be compact or of irregular plan. In the Dinaric karst the small Blato polje is practically as wide as it is long but the Slamoc polje is 26 times as long as it is wide. The largest is the Livanjsko in Croatia which is about 40 km long and 6-8 km wide. Gams (1969) requires the flat floor to be

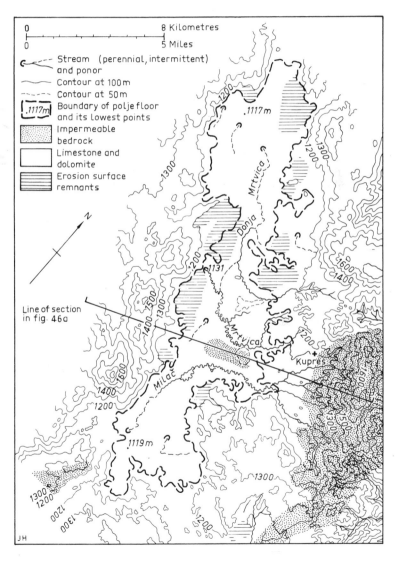

44 Kupres polje, Bosnia. After Roglić 1939.

24 Planina Polje, Slovenia, under winter floods which recur for several weeks each year. Photo by I. Gams.

25 Small polje in Napier Range, West Kimberley, Australia. Alluviated area beneath woodland.

several square kilometres in area before a closed depression qualifies as a polje.

There is generally a sharp break of slope between the floor and the sides, which are normally fairly steep. Where these sides are in impervious rocks, streams flow down on to the floors; elsewhere they rise in springs at their margins. Then they flow over alluvial fans or plains to low points where they sink. These streamsinks are called ponors, which vary from cave entrances in the limestone walls to alluvial streamsink dolines. There may be several ponors along a stream or else it may branch to several ponors. There may be flood ponors and old abandoned ponors at levels higher than the normal place of engulfment. Small poljes may have only one alluvial floor and one stream; others have several, with low, doline-riddled bedrock areas between.

In many poljes, the ponors cannot carry away the runoff fast enough after rainy weather, even when higher-lying ones have come into action. Then shallow lakes form over part or even the whole of the polje floor. Thus Popovo Polje, inland from Dubrovnik in Dalmatia, was inundated over three-quarters of its floor in each of seven years between 1891 and 1900, with depths of more than 30 m at its lower end. In some poljes certain ponors change function for a period in the wet season and spew out water instead of imbibing it. Indeed this is the commonest context for estavelles.

During the dry season temporary lakes recede as runoff fails and evaporation returns water to the atmosphere. There are, however, permanent lakes over parts of some polje floors, e.g. in Cerkniča polje, Slovenia. Near coasts, lowlying poljes are sometimes entirely occupied by permanent lakes, e.g. Jesero Lake near Zara in Dalmatia.

In the Dinaric karst and elsewhere there are depressions fully comparable with poljes except that they have external surface drainage along a narrow defile which seems almost adventitious to the major feature. Cvijić (1960) terms them *open poljes*; semiclosed may be a better term.

Residual hills of limestone or *hums* in some cases protrude through the alluvial plains or rise from slightly higher bedrock floors of limestone, particularly towards the sides of the poljes. In Yugoslavia they vary from 15 to 90 m in height and have a sharp break of slope at their foot. They are often pyramidal in shape with rather uniform slopes but those in the alluvium tend to more convex profiles with a basal steepening of slope (Klaer 1957).

Early debate on the formation of the Yugoslavian poljes has been confused by the facts that many are closely related to the mid-Tertiary folding of the Dinaric mountains and that late Tertiary lacustrine beds occur in some of them. This led to the idea that they are of tectonic origin, occupying fault-angle depressions, fault troughs and synclines (Fig. 45a). There can be little doubt that some poljes constitute primary tectonic relief.

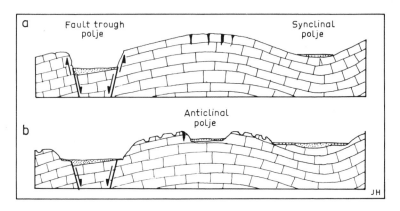

45 *Structural relations of poljes*
   (a) Primary tectonic poljes in syncline and fault trough.
   (b) Enlargement of tectonic poljes and development of secondary poljes by corrosion.

Thus in New Guinea, the depression in which Lake Tibera lies is a tectonic depression due to high angle reverse faulting in a Plio-Pleistocene orogeny, subterraneously drained through limestone. However, the Dinaric karst is older and there are thought to be high-lying erosion surfaces which argue against survival of tectonic relief. The Tertiary lacustrine beds have been tectonically deformed themselves and are older than the time of polje formation. Some poljes are formed in tectonic highs, namely anticlines and horsts. Moreover poljes in synclines and fault troughs often have irregular margins which transgress the faultlines and synclinal limbs broadly defining the polje floor. Additionally these floors, whether exposed or alluviated, are found sometimes to truncate structures in the limestone (Fig. 45b). Therefore in the classical Dinaric karst the close relationship of poljes to tectonic lines seems to be a reflection of structural guidance of erosion. Even the simplest tectonic poljes usually involve some erosion, e.g. Mende polje, Portugal (Birot 1949; Fig. 46c).

Erosional poljes in Yugoslavia have been interpreted by Roglić (1939, 1940) as due to lateral solution undercutting, a mechanism applied to various areas by others (e.g. in Turkey by Louis 1956;.

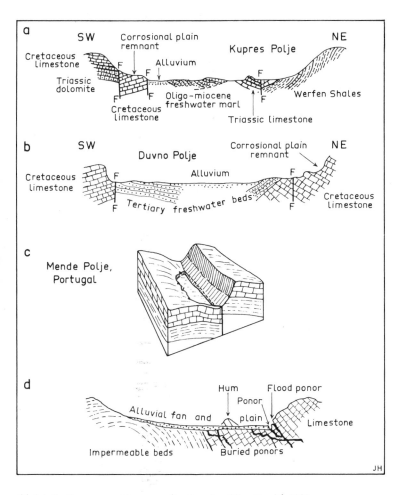

46 (a) *Section across Kupres polje, Bosnia. After Roglić 1939.*

(b) *Sections across Duvno polje, Bosnia. After Roglić 1940.*

(c) *Block diagram of Mende polje, Portugal. After Birot 1949.*

(d) *Theoretical section to show polje formation by lateral corrosion from alluvial floor. Based on Roglić and Louis.*

in Italy by Lehmann 1959). Parts of the walls and floors of the poljes Roglić studied are developed on impervious Mesozoic and Tertiary beds and this is where they initiated (Fig. 46a, b). Surface drainage on them carried detritus to streamsinks in the surrounding limestone where alluvium accumulated (Fig. 46d). Because of its lower solubility, dolomite has acted similarly so that some poljes have developed along the contact of limestone and dolomite (Gams 1969). Periodically the surface of the alluvial flat formed in this way was flooded and aggressive waters attacked the limestone margins. Rotting organic matter washed to these contacts provided biogenic CO_2. By this biochemical solution at the edge of the plain, a flat floor was extended into the limestone and an alluvial seal simultaneously spread over it. This seal normally protected the karst rock beneath, permitting this extension by lateral solution. Poljes created in this way can obviously be regarded as developments of blind valleys.

At certain junctures, ponors may open up substantially, enabling polje streams to incise the alluvium; subsequent partial blocking of the ponors would result in fresh lateral planation at a lower level, producing a countersunk floor. Rock terraces along polje floor margins occur which could correspond with such a history.

Polje floors represent interior planation surfaces. H. Lehmann and others have stressed control of their altitudes by the level of the outflows *(Vorfluters)* of the underground drains; this can result in common levels for many poljes where the outlets are controlled by sea level or that of a coastal plain. The other school of thought led by Roglić emphasises control by ponor level within the poljes whereby each polje, even parts of poljes, are independent in their planation level. The cave systems between ponor and outflow vary so much that there is good reason to think that both circumstances occur in nature.

From Cuban tropical humid karst Lehmann and others (1956) described karst margin poljes where the flat floor is largely surrounded by impervious rocks with a limestone wall on the outflow flank only. Panoš and Štelcl (1968) have denied that much of these floors is developed on former limestone floors subsequently stripped, as Lehmann inferred. There is little doubt but that this has happened in the case of a karst margin polje behind The Tunnel in the Napier Range, West Kimberley, where the backreef facies of a Devonian reef structure has largely been lost in this way. Karst margin poljes have been recognised in many karsts. However, the term polje may not strictly apply to closed depressions without any limestone floor in their evolution. Some of the 'inland valleys' surrounded by limestone in Jamaica but entirely floored by impervious bedrock are of this type.

Though the Yugoslavian terms—doline, uvala, and polje—can be used to classify very many closed depressions in karst, there is greater variety than this and no attempt should be made to force all occurrences into this framework.

KARST MARGIN PLAINS

Attention has concentrated on the flat surfaces which have developed enclosed within karst because they are the most distinctive. However, many of the factors which operate there can also be effective in the condition of external surface drainage. Morawetz (1967) has redrawn attention to the lower Neretva valley in Yugoslavia, a plain only 4-6 m above sea level as much as 20-30 km inland. It is a bedrock plain with an alluvial veneer, hums project through it and the margins are embayed by steepheads. This active *karst margin plain* is being extended today by spring sapping and also by general flooding in winter and spring from the Neretva River and the springs. Corrosion plains of this type tend to greater perfection than most other kinds of degradational plain. Jennings and Sweeting (1963) have argued that tropical semiarid pediments likewise tend to a higher degree of planation as a result of special processes obtaining when they are developed on karst rocks. But it is in the humid tropical karsts that the best examples of karst margin plains are to be found, for example in the Celebes (Sunartadirdja and Lehmann 1960) and in Tabasco, Mexico (Gerstenhauer 1960) because the climate and vegetation are there very favourable to extension of plains by lateral solution (see Chapter IX). The extensive opencast workings for tin in the plains of the Kinta valley in Malaya reveal vividly how intricately corroded subsurface relief with an amplitude up to 10 m is nevertheless horizontally truncated most sharply.

VII

KARST CAVES

Penetrable natural cavities are formed in diverse ways—by lava flow, wave attack, weathering, landslides, and movement and melting of glaciers, but caves formed by karst processes are the most numerous, largest and most complex. Different karst rocks vary in their propensity to form caves, chiefly according to their chemical purity and mechanical strength. Insoluble residues tend to block incipient cave development and inadequate shear strength results in collapse of developing cavities. The last factor explains why caves are poorly developed in evaporites (Krejčí-Graf 1935; Wagner 1935; Würm 1953). Rock strength increases with compaction, itself the result of compression, cementation, and recrystallisation, but decreases with frequency of planes of weakness. Some planes of weakness are necessary for substantial permeability, which is another prior condition for cave formation. Yet no karst rocks seem too massive for caves, though they may be few and large in very massive rocks. Coarse, intergranular porosity may maintain dispersed percolation and be inimical to speleogenesis, the formation of caves; this may explain the comparative poverty in caves of the Chalk of Britain and France. Water is also necessary for caves to form. The Nullarbor Plain is possibly the largest continuous limestone area in the world, a free karst of approximately 200,000 km², yet cave development is modest because it is entirely subject to semiarid and arid climates (Jennings 1967a). Where rock temperatures are below freezing point, water is present as ice only and caves cannot develop, though they may survive (Corbel 1954).

Though predominantly the effect of meteoric waters, solution of karst rocks is not quite exclusively so. Thus hydrothermal water produces caves and cave decorations and these must be considered true karst features. One of the best known occurrences of hydro-thermal karst is at Hranice on the River Becva in Moravia (Kunský 1958).

GENERAL CHARACTERISTICS

Karst caves are extremely varied in all respects. They range in length from a few metres to well over 100 km in the cases of Flint Ridge Cave, Kentucky, and Hölloch in Switzerland and in depth to 1311 m in Gouffre St Pierre Martin in the French Pyrenees. Calculations of area and volume are few so the range cannot be specified, but very large caves in these two ways are the Hölloch with 7 km^2 of area and Gouffre Berger in the French Alps with 2-3 km^3 of volume (Gèze 1965). In complexity they vary from single rooms, short passages, or open shafts (potholes) to most intricate systems of passages linking rooms, shafts, chambers, and halls of all sorts of shapes and sizes. Single chambers reach the size of Grotta Gigante near Trieste, which is 200 m long, 130 m wide, and 136 m high, and of the Big Room in Carlsbad Cavern, New Mexico, with dimensions of 400 m, 230 m, and 100 m. Associated with penetrable caves, there are usually innumerable enlarged joints and small tubes, embryonic caves, which together may amount to a much greater volume than the caves proper. The density of caves can vary from a few m/km^2 in several small caves to hundreds of metres of cave per square kilometre. Some residual hills of limestone seem to be mere shells riddled with large and small cavities, e.g. Mt Etna near Rockhampton, Queensland, though more sober examination proves that much more rock remains than voids.

Some caves are completely dry, others are filled with water. The Masocha pothole in Moravia has a lake more than 100 m deep and there are more than 10 km of lakes (more like canals than true lakes) in Padirac in the Causse de Gramat, central France. Some caves are completely surfaced by bedrock but most have deposits of various kinds in them, which can accumulate to total fill when the cave becomes fossil.

Caves occur in all kinds of topographic situation from beneath coastal plains and valley bottoms to the tops of mountains. High, steep ridges may be poor in caves because gradients have favoured runoff at the expense of infiltration (Grund 1910b). High plateaux tend to be rich in deep potholes and caves with a great deal of vertical development whereas low plateaux will chiefly have horizontally developed caves. Cave exploration beneath lowlying plains and valley bottoms usually involves diving, so not much is known about them in this situation.

Cave entrances may be vertical shafts and fissures or lateral openings in slopes and cliffs. When small, the aperture may have smooth forms and be entirely erosional, but when large, rockfall has usually contributed to their enlargement. Even then they need not be angular because subaerial weathering can round off large arches produced by collapse. Tiny entrances can lead into large systems and many caves have only artificial entrances, usually made by removal of rockfall or digging out of finer detritus. Some caves have only become known when quarrying or mining in bedrock intersected them (Warwick 1968). Other caves have water-filled entrances.

Very many caves have rivers running through them and are called active river caves. These caves can usefully be considered on the following basis (Grund 1910b):

(a) *Inflow cave* where a river is followed downstream from its point of engulfment to a sump which is the limit of exploration.

(b) *Outflow cave* where a river is followed upstream to impassable obstacles or to the beginnings of concentration of seepage water.

(c) *Through-cave* where a river is passable from its sinking point to its rising, possibly through water-filled sectors.

(d) *Between-cave* where a river passage is entered from above or laterally but cannot be followed to the surface either upstream or downstream.

The term 'dead cave' is usually applied to caves which are not being enlarged by water action today; many of these have forms and deposits which reveal they were active river caves in the past. The simple classification above can be applied to these also. Drainage through karst rocks tends to develop lower routes to the risings. This happens both through self-capture and through piracy of neighbouring cave rivers. River caves can become completely abandoned in these ways. Often cave systems consist of abandoned levels at successively lower altitudes, with the lowest one remaining active (Fig. 47).

In caves, there is an almost infinite variety of bedrock forms both large and small, which result from the interaction of active processes with the passive factors of rock types and arrangement (Bögli 1956; Renault 1958). These are called *speleogens* and some of them will be mentioned where appropriate in the following discussion.

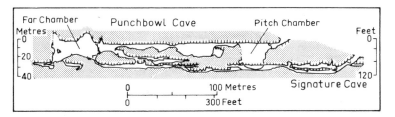

47 Long profile of Punchbowl-Signature Caves, Wee Jasper, N.S.W., with four levels of development and subhorizontal solution roofs (hatched). Bedding is vertical.

CAVE FORMATION

It is impossible in this small book chiefly concerned with surface geomorphology to do justice to the great controversies of the past about cave formation or even to the manifold views held today by speleologists about the caves with which they are individually familiar. Only the main threads can be followed.

Tectonic caves

Tectonic origins for karst caves lost their early appeal during the gradual overthrow of catastrophism by uniformitarianism in the explanation of landforms. The great numbers of caves in karst rocks invalidate such origins as a general cause of these caves because tectonism would affect all rocks of comparable strength to much the same degree. Associations of caves and tectonic features such as the slickensided fault-plane roof of Terrace Chamber in Marakoopa Cave, Mole Creek, Tasmania, generally imply no more than structural guidance of the processes forming caves. Nevertheless some caves may be due directly to earth movements as Gèze (1953) maintains for some potholes of fissure type. The 90 m deep Igue des Landes in Tarn, France, is due to the opening up of joints in massive limestone in an anticline, whereas the Avens du Pic St-Loup near Montpellier are in vertically dipping beds, which have gaped apart through spread under gravity (Fig. 48a). It is in Cainozoic orogenic belts that there is the greatest likelihood of some caves being due to earth movement; thus some of the large caverns in the Dachstein plateau in Austria have been attributed by Arnberger to differential movements of rock masses under thrust faulting (cf. Groom and Coleman 1958; Fig. 48b). It is postulated that these were later joined to one another and to the surface by passages of erosional origin.

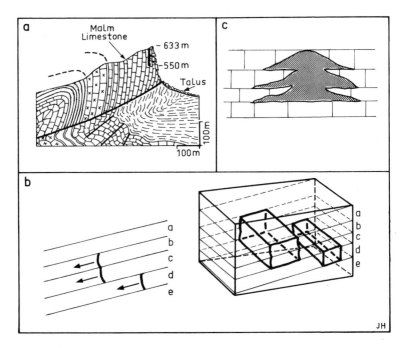

48 (a) *Avens du Pic St-Loup, Montpellier, France, potholes due to gravity spread of vertical beds. After Gèze 1953.*
   (b) *Arnberger's concept of tectonic cave formation by differential movement of limestone bodies along bedding planes.*
   (c) *Shelves in walls of caves, Clare, Ireland, through differential solution of limestone beds. After Ollier and Tratman 1969.*

Renault (1967) has argued that residual tectonic stresses released by both surface and underground erosion play a significant role in the localisation and growth of caves.

Corrasion versus corrosion

By the late nineteenth century the central controversy in speleo-genesis had shifted to the relative roles of corrasion and corrosion in the erosional work of underground rivers.[1] Cave explorers have understandably been impressed by the noise of waterfalls and the grinding of boulders together in underground confines where even small rapids and whirlpools can present real obstacles to move-ment. There resulted a tendency to attribute more or less all the work in forming caves to the enlargement and fusion of fluvial potholes or rock mills. It was necessary for the case to be argued for the less obvious quiet chemical action of underground water. In extensive karsts, underground rivers often lack boulders, pebbles, and sand for corrasional attack, precisely because the rock fragments are dissolved, though with sufficient velocity purely hydraulic action remains in their power. But the speed of move-ment required with or without rock tools presupposes a fairly large passage, which must first be produced by solution. Where on cave surfaces there is differential solution of different limestone beds or of chert inclusions, the dominance over mechanical attack is also proven (Ollier and Tratman 1969; Fig. 48c).

Indeed appreciation of such facts and concentration on the distinctive karst process of solution have led in recent decades to undervaluation of mechanical attack. But small impounded karsts usually have their underground rivers fed with detritus from surrounding impervious rocks. For example, on Cooleman Plain, N.S.W., nearly all caves have trains of igneous pebbles through them from surrounding ranges, even though they are practically horizontal systems. The more extensive limestone areas of the King Country, New Zealand, are interspersed with outcrops of other sediments and there have been extensive ignimbrite and volcanic ash covers. Signs of powerful mechanical action such as plunge pools and rock mills are very evident in the caves of this high rainfall area. High mountain karsts reaching to frost wedging levels have good cave supplies of rock fragments fed down pot-holes and streamsinks into their caves.

Martel (1921) rightly stressed that the two aspects of erosion

---

[1] It should be noted that in much French and German literature corrasion and erosion are equated.

26 *Cave river waterfall and plunge pool, The Chute, Coolagh River Cave, Clare, Ireland. Water about 70 cm below normal level. Photo by D. M. M. Thomson.*

generally accompany one another but that their relative roles vary with lithology and hydrodynamic circumstances.

Vadose zone action

The conditions governing cave forming processes vary between different parts of a karst hydrologic system. At one time or another speleogenetic theory has stressed each of the hydrodynamic and chemical contexts which occur in these systems. Stress was first of all laid on the action of seepage water and free surface streams in the vadose zone because this was the most readily examined. In some ways matching grikes and potholes on the surface, there are blind fissures and shafts rising from cave passages and chambers towards the surface but not reaching it. Some blind shafts[2] and some potholes are attributable to seepage water, which undoubtedly seams the walls of many of them with vertical grooves separated by thin, sharp ribs. Cylindrical vertical shafts of the central Kentucky caves in Pennsylvanian limestone are significantly located beneath the edges of sandstone-capped plateaux at the heads of recesses. According to Pohl (1955) they are fortuitously related to the horizontal cave systems, which themselves disregard the surface topography. He argues therefore that these blind shafts are due to solution down joint intersections by aggressive seepage water percolating through the sandstone (Fig. 49a). Burke and Bird (1966) have similarly interpreted closely comparable forms in the Carboniferous Limestone beneath the Millstone Grit margin in South Wales; the Gunbarrel and other blind shafts in Wyanbene Cave, N.S.W., stoped to the unconformity between Devonian conglomerate and Silurian limestone, provide another match.

Very much larger potholes, often of inverted conical form, such as Gouffre Henne Morte (466 m) in the French Pyrenees, are attributable to seepage water since they are fed only by small doline catchments in mountain tops (Fig. 49b, Pl. 33). Something

[2] Terminology for blind shafts in English is not very satisfactory. The British have long misused the French *aven* for them but in France *aven* means a shaft open to the sky, a pothole, in fact. The Americans use the Italian *foiba* (pl.-*e*) but a similar twisting of meaning on transfer seems to apply here also. The American 'dome pit' is also not satisfactory for general use because many blind shafts are not domed at the top but taper off. Blind shaft seems short and specific enough to act as a generic term and is therefore used here.

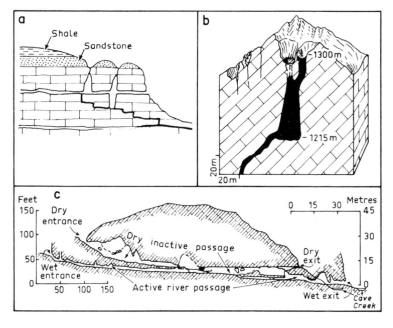

49 (a) Vertical blind shafts or foibe in limestone caves beneath sandstone and behind scarp recesses, Kentucky. After Pohl 1955.

(b) Upper part of Gouffre Henne Morte, Pyrenees, France, pothole due to seepage from small surface hollow. After Gèze 1953.

(c) Long profile of Barber Cave, Cooleman Plain, N.S.W. Dry passage shows river concavity of upper vadose section and subhorizontal solution roof of lower shallow phreatic passage.

of the role of a sandstone capping not present here may be taken on by snow drifting into the dolines in great banks which release water slowly. Gèze (1953) calls them absorption potholes. He and others such as Maucci (1960) maintain they grow upwards though the water is descending. Water condensing on cold rock surfaces in the vadose zone may in some circumstances significantly contribute to the formation of these potholes.

Similar potholes are caused by streams sinking vertically after passing on to karst rock from a catchment on impervious rocks, which may overlie it or be faulted against it. There must be deep valleys in or around the karst for deep potholes. One of the deepest potholes in Britain, Gaping Gill Hole in Craven, was formed in Carboniferous Limestone by a stream from the overlying and mainly impervious Yoredale Beds which include much shale. Only flood waters of Fell Beck nowadays fall down the pot; its normal discharge sinks a little upstream and enters it below the surface. As the boundary of the Yoredale Beds retreats under erosion, fresh streamsinks are created and potholes of this origin are abandoned. Bar Pot appears to be an inactive predecessor of Gaping Gill Hole with its entrance blocked by rockfall, recently opened artificially.

Another kind of pothole is the consequence of the ancillary process of collapse also characteristic of the vadose zone. Successive breaking away or stoping of a cave roof eventually penetrates to the surface. A block pile or debris cone accumulates below (subject to removal by river action). The Daylight Hole in the Dip Cave, Wee Jasper, N.S.W., has a block pile and is situated on the crest of a spur so that it is unlikely to have been a streamsink. However, rockfall can contribute to the growth of seepage or streamsink potholes so there are transitional forms just as enlargement of collapse potholes produces collapse dolines.

In the vadose zone, free-surface streams have downhill gradients under gravity, sometimes developing an exponentially concave longitudinal profile as do surface streams (Fig. 49c). Plunge pools occur below waterfalls and rock mills are found in the bottoms of canyons, the walls of which sometimes show remnants of the cylindrical pits drilled by them (Ford 1965a). Where gradients are less, the walls may still give evidence of vertical downcutting in sequences of nearly horizontal stream grooves (Pl. 28). Lateral erosion accompanying the incision may result in ingrown meanders with the roof matching the slip-off slope. Meander spurs may be undercut and left suspended (Jennings 1964). Alternatively meanders may be cut off and oxbow passages left abandoned. In plan the confluences of tributaries produce branchwork patterns (Fig. 50a), but the confined nature of caves, with no equivalent of the flood plain on the surface, results in ready rise of level in floods and there is a strong tendency for oxbows and higher level passages normally left dry through capture to be maintained in periodic action as flood overflow routes. In some caves such as Honeycomb Cave, Mole Creek, Tasmania, the river in flood flows through channels which divide and rejoin complexly, not in one plane but several; it is in some degree like a three dimensional braiding system. Structural lineaments—joints, bedding planes, faults—often strongly influence the plans of vadose caves but gravity stream action will frequently impose independent curving elements.

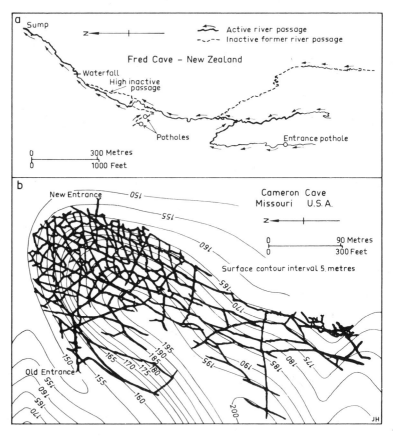

50 (a) *Plan of Fred Cave, Auckland, N.Z. Between-cave of mainly vadose development with branchwork pattern influenced by joints. After N.Z. Spel. Soc. survey.*

(b) *Plan of Cameron Cave, Missouri, U.S.A. Two-dimensional network or maze developed phreatically with joint control. After Missouri Spel. Soc. survey.*

Microforms consequent on turbulent flow in the vadose zone are current markings or scallops (Fig. 51), asymmetric hollows with a steep semi-circular step on the upstream side and a gentle rise downstream ending in a point between the steps of the next downstream hollows (Coleman 1945; Ollier and Tratman 1969). Friction with the surface causes small eddies to develop which alternate with laminar flow; solution is greater in the eddies and so hollows the rock. The slower the flow the larger the scallops (Glennie

1963). Flutes (Curl 1966a) are similarly asymmetrical solution hollows but with long parallel crests transverse to the current. Curl has developed a mathematical function relating their dimensions to the hydrodynamics of the flow.

Scallops and related forms are caused by solution, but in the vadose zone it is possible that corrasion may match or even surpass corrosion in enlarging caves in some circumstances.

Deep phreatic solution

A free-surface stream presupposes a cave passage through which it can move freely. In a paper stimulating much rethinking about cave formation, Davis (1930) argued that most of these initial passages were inherited from a previous phase of solution by slow moving water in the phreatic zone below the watertable. Such water infiltrated along the planes of weakness, enlarging them by

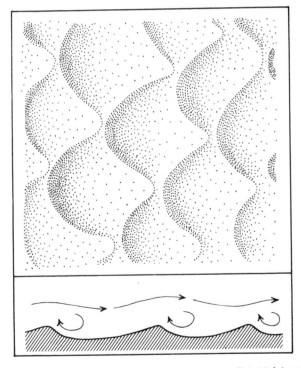

51 *Current markings or scallops developed from small initial irregularities by eddy currents*

solution into three-dimensional networks of small galleries and shafts, irregularly connected, with outgoing as well as incoming branches and with rising as well as falling gradients. These phreatic caves would develop best beneath erosional planes truncating the limestones. Uplift and dissection then resulted in partial drainage and modification of the caves into branchwork systems, with corrasion and breakdown added to solution as sculpturing processes. The very close adaptation to structure of the phreatic caves would be partially destroyed at this time.

Bretz (1942) elaborated and gave precision to these ideas. The elaboration included the insertion of a red clay infilling between the phreatic and the vadose phases. Though such a history may be true of particular caves, many cave studies over the world show that it is not usual to have a stage of red clay fill. On the other hand much of Bretz's discussion of criteria to distinguish between vadose and phreatic origins for caves has proved valid and useful.

Thus substantial development of the following features is interpreted as due to solutional work in the saturated zone without definite currents:

(a) *Spongework*; intricate cavitation like the pores of a sponge on various scales (Pl. 27).

27 *Phreatic spongework in Tunnel Cave, Borenore, N.S.W.*

(b) *Bedding plane and joint anastomoses*; repeatedly branching and joining patterns of small sinuous tubes in such planes. With bedding planes, the holes of the anastomoses often have flat bottoms in them, residues sealing off the lower bed.

(c) *Wall and ceiling pockets*; hemispheroidal hollows in these surfaces of passages and chambers. Bögli (1964a) attributes ceiling pockets or 'bellholes' along joints to mixing corrosion.

(d) *Joint wall and ceiling cavities*; similar but elongated along joints.

(e) *Ceiling half-tubes*; larger features than (b) in roofs of large chambers.

(f) *Continuous rock spans* across cave chambers; *bridges* if more or less horizontal, *partitions* or blades if more or less vertical.

(g) *Two- and three-dimensional networks* or mazes of passages (Fig. 50b); once created, these can be maintained and enlarged in other hydrodynamic states.

Rock pendants are bedrock projections hanging from ceilings, smoothed erosionally and often in groups reaching down to a common level. Bretz attributes them to vadose modification of the rock left between anastomosing half-tubes, but others consider them completely due to some kind of phreatic solution.

Most caves exhibit some of these characteristics and true phreatic solution must initiate most cave development. However, substantial enlargement in deep phreatic conditions beyond this phase is thought to apply only to particular cases. Dip Cave at Wee Jasper, N.S.W., appears to be a little modified phreatic system with quite large chambers; breakdown since emptying is the chief modifier. But clay residues will be susceptible to flocculation in the alkaline water and will settle out as circulation is slow. Diving into water-filled caves has revealed that clay coatings are common on walls, floors and ceiling, which must hinder further solution of these rock surfaces.

Watertable stream, shallow phreatic and pressure passage action

If one tube in a bedding-plane anastomosis or one element in a spongework gets larger than its neighbours, water will pass through it more rapidly and enlarge it still further. This autocatalysis changes hydrodynamic conditions substantially and is most likely to occur at the top of the watertable or rest level where accessions of fresh water may renew aggressiveness.

Considerations such as these and the other restraints on deep phreatic solution cited above led Swinnerton (1932) to maintain that most solution went on near the top of the permanent phreatic zone and in the intermediate zone through which the watertable oscillates. Fast turbulent watertable streams fashion big subhorizontal tunnels disregarding structure. Roofs as well as walls and floors are subjected to solution so that asymmetrical current markings can be found on ceilings also, though smooth surfaces can occur in these tunnels as well. Flat ceilings cutting across structures (Fig. 47, Pl. 28), have been regarded as diagnostic of action at this level by several investigators (Halliday 1957; Jennings 1964). Lange (1962) discusses the related question as to how ceilings are planed off by solution at regulatory water levels of cave lakes.

28 *Dry passage in Barber Cave, Cooleman Plain, N.S.W., with shallow phreatic flat roof, vadose current markings, and successive channel grooves in walls*

Most cave investigation in the last two decades has confirmed Swinnerton's views rather than Bretz's, though the term *shallow phreatic* has been preferred for this important zone of action (Davies 1960; Deike 1960; White 1960; Thrailkill 1960, 1968). The *master caves* of Craven seem closely comparable to the shallow phreatic type (Glennie 1954b) and Glennie's term *epiphreatic* is also commonly used for this hydrodynamic zone (Glennie 1958). Where caves are developed in gently dipping beds which vary in solubility, abrupt terminations of passages are regarded by White (1960) as pointing to watertable control. Successive horizontal levels boring through steeply dipping beds and related to surface river terraces (Davies 1960; Ek 1961; Fig. 52) and the association of major cave development in karst areas

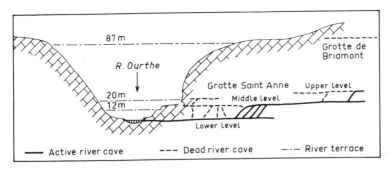

52 *Relationship of cave levels in Grotte de Briamant and Grotte Saint Anne at Tiliff, Belgium, to terraces of River Ourthe. After Ek 1961.*

with the major valleys are also interpreted in the same way. Pitty (1968) quotes various groundwater engineers to the effect that there is a rapid decrease downwards in limestone of cavities and of rate of water circulation. The chemical characteristics of the top of the phreas with several modes of renewal of aggressiveness may be as important as the kinetic considerations already set out here.

On the continent of Europe where there are frequently great depths of cave and large inputs of water and where the notion of watertables in most karst is often unaccepted, speleologists have chiefly contrasted pressure flow and free surface flow (Chevalier 1944). In the parts of caves permanently or temporarily filled with water, hydrostatic heads can build up through inadequate capacity to cope with increases in discharge; solution and corrasion can then affect all surfaces of passages in rising as well as falling passages. Both true and inverted siphons come into violent action.

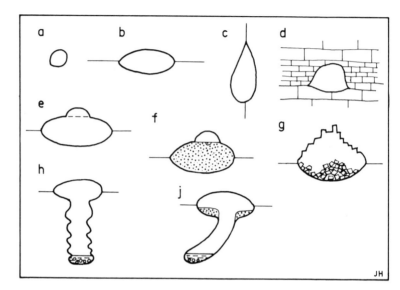

53 *Types of passage cross-section. After Renault 1958, Bögli 1956, Ollier and Tratman 1969.*

   (a) *Phreatic tube in massive rock.*

   (b) *Elliptical phreatic passage in horizontal bedding plane.*

   (c) *Phreatic passage in vertical joint plane.*

   (d) *Phreatic passage in group of more soluble beds.*

   (e) *Phreatic passage with ceiling half-tube due to air entrainment along roofline.*

   (f) *Phreatic passage aggraded to roof with development of ceiling half-tube.*

   (g) *Elliptical phreatic passage modified by breakdown.*

   (h) *Vadose canyon with horizontal channel grooves incised in floor of phreatic passage.*

   (j) *Vadose canyon with inward meandering. Some aggradation of primary phreatic passage before incision.*

Pressure passages of this type can assume various forms (Renault 1958) according to structural constraints, the distinctive criterion being liability of all surfaces to erosion (Fig. 53a-d). Cylindrical tubes develop where there is little structural influence (Pl. 29); vertical or horizontal elliptical passages where planes of weakness retain control; rectangular passages where beds differ markedly in their response to erosion. A ceiling half-tube (*Wirbelkanal:* Bögli 1956; Fig. 53e) may result from air entrainment accentuating solution at the highest roofline (Pl. 30). These pressure passages may be modified subsequently by the introduction of fill, vadose flow or collapse. Some apparent tubes are not due to pressure flow at all but to negative exfoliation due to rock

29 *Keyhole passage in Metro Cave, Charleston, New Zealand. Upper part phreatic with vadose incision in floor. Photo by D. L. Homer by courtesy of New Zealand Geological Survey.*

30 *Speleothems in Metro Cave, Charleston, New Zealand. Straw and conical stalactites, stalagmites, and a column. Part of a shawl at top left. Photo by D. L. Homer by courtesy of New Zealand Geological Survey.*

pressure (Fig. 6a). Signs of rock spalling will distinguish these from fresh pressure tubes but the solution surfaces of long abandoned pressure tubes will come to approximate to the condition of those due to mechanical breakdown alone.

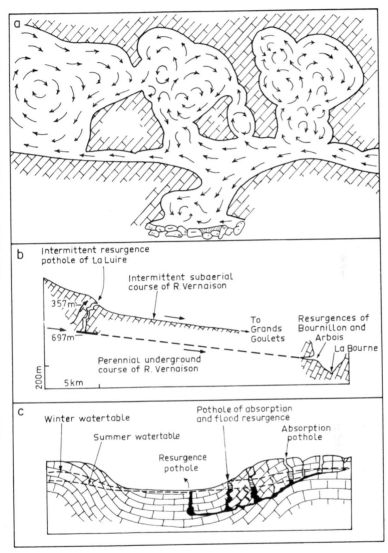

54 (a) *Symmetrical solution hollowing of Picznice Cave, Hungary. After Cramer 1933.*

  (b) *Gouffre de la Luire, Vercors, France, resurgence pothole. After Gèze 1953.*

  (c) *Potholes round margins of polje, Grand Plan de Canjuers, Provence, France. After Gèze 1953.*

Forms intermediate in nature between those of deep and shallow phreatic or pressure flow have been recognised. Injections of powerful currents into large water-filled spaces seem capable of producing large symmetrical surface hollows. Cramer (1933) describes this from the Piznicehöhle in Hungary (Fig. 54a), and similar sculpture is characteristic of Cathedral Cave, Wellington, N.S.W. Small symmetrical hollows transitional to scallops are also known, e.g. in Dip Cave, Wee Jasper, N.S.W. (Jennings 1963).

Pressure flow is involved in two kinds of pothole (Gèze 1953). Resurgence potholes are essentially due to such flow. The Gouffre de la Luire in the Vercors plateau, French Alps, is perhaps the most striking example; hydrostatic pressure forces water more or less vertically up 200 m from a deep river cave when it is in flood (Fig. 54b). High ground above is necessary for the pressure required and usually a fault or a syncline provides favourable structural conditions. Other potholes are due to alternations of rising pressure flow and descending gravity flow according to changing water levels. These occur round the margins of poljes, e.g. the estavelles around the Grand Plan de Canjuers in Provence (Fig. 54c).

As a karst develops, potholes may change their function permanently, so each one needs careful analysis. Indeed a catholic attitude towards hypotheses of origin for all caves is necessary because most caves are composite in their nature. Ascertaining the relative importance of each kind of action which has produced the present form and pattern of each cave depends on reconstructing its history. In many parts of the world the conclusion has been reached that the most diverse histories may be expected from caves quite close together, e.g. in the Sierra Nevada (Halliday 1957, 1960), the Mendips (Ford 1965b), and Wee Jasper, N.S.W. (Jennings 1967c).

Large chambers

Large chambers in cave systems are often difficult to understand because much of the evidence is destroyed in their formation (Renault 1967). Such chambers may be located excentrically with respect to the main passages but often they lie at the junction of several passages. Sometimes favourable structural factors may be discerned, e.g. faults, close-set joint fields, lithological weaknesses. The Grotte de la Cigalère in the French Pyrenees has chambers wherever the cave encounters schists encompassing the limestone in which it is found.

Low, wide chambers may be due to solutional removal of blocks between a labyrinth of small passages in a single bed. This seems to apply to Flat Roof Lake Chamber, Jewel Cave, Augusta, southwestern Australia, which is 160 m long and up to 45 m wide, and developed in a shallow phreatic zone. When a roof gets too wide to support itself, collapse results in a domed chamber. The deep caves of the Nullarbor Plain have many such which stope through weak chalky limestone of the Eocene Wilson Bluff Formation to the stronger overlying Miocene Nullarbor Limestone. Other chambers seem to be formed from labyrinths at several levels above one another when collapse will contribute as well as solution. This is the explanation given for the large halls of Mullamullang Cave in the Nullarbor Plain by Hunt (1970). Rockpiles result from such developments.

Most of the chambers of Punchbowl Cave, Wee Jasper, N.S.W., occur where the stream forming the cave stayed in approximately the same position during the time more than one of its four levels formed (Jennings 1964); only in the Far Chamber is there collapse material and this has entered laterally. The ceilings of the chambers are mainly of solutional nature, e.g. Pitch Chamber. Renault (1958) attributes the Grotta Gigante, which is as deep as it is wide and not a great deal longer, to collapse associated with pothole development; there are three pothole entrances and the rockfall floor is hollowed. It is an 'underground doline' and there must have been removal from below after collapse.

MORPHOMETRIC ANALYSIS OF CAVES

Caves offer abundant scope for morphometric analysis of both large and small forms, yet not much has so far been essayed in this field, partly because underground conditions are often not conducive to tedious repetitive measurement. The commonest exercise has been to analyse the directions of approximately straight passage segments and relate these to structural controls such as strike joints, dip joints, and (in inclined rocks) bedding planes (e.g. Gams 1963). Glennie (1948, 1950) found that in Ogof Ffynnon Ddu in South Wales the frequency maxima of joint-aligned passages coincided with the strike and the dip in the eastern part of that cave but in the western part they occur 20° away from both these directions. He surmised there may have been earth movement since the latter part of the cave formed.

Less frequent has been analysis of the attitude of planes of weakness which have governed cave cross-sections. However, Maucci (1960) has done this for a sample of 200 caves in sub-horizontal limestone near Trieste. Not unexpectedly more than half were vertically disposed or close to it, nearly half lay between the horizontal and 30° from it, leaving only a very small number between 30 and 70°. However, with such figures he rein-forced his arguments that his 'orthovacuums' (i.e. blind shafts and potholes) have quite separate genesis from his 'paravacuums' (i.e. caves proper), the first being dominantly due to seepage and the latter to streamflow. He maintains that because of gravity, seepage solution is most effective in vertical planes of weakness, producing the dominant frequency about the vertical. Comparison with similar data from karst with about a 45° dip would be interesting.

These examples used simple analysis for descriptive purposes only. Employing established statistical theory for immigration-emigration processes based on random walks, Curl (1958) derived a stochastic function for the relationship between lengths of caves and the number of entrances they have, on the assump-tions that the longer a cave the more likely it is to gain additional entrances by natural processes such as solution and the more entrances a cave has the more it is likely to lose some by other natural processes such as collapse. Applying this to limestone caves from West Virginia and Pennsylvania, he got good agree-ment between observed and predicted length/entrance data.

For the West Virginian caves, Curl (1960) selected a mathe-matical model for changing length of caves with single entrances based on a regional history of a phreatic phase in which there was increase in length of a fixed number of caves, of a vadose phase after rejuvenation of cave fragmentation and length reduction, and of a final decay phase of length reduction without loss of cave numbers. This was applied to the observed length data and past and future length frequencies predicted.

Later Curl (1966b) applied the same approach to a greater range of limestone karsts for which appropriate data were avail-able. The assumption that the number of cave entrances is related in a simple direct manner to cave length was tested against two other assumptions and was found the most acceptable statistically. The preferred model predicted a considerable number of entrance-less caves, on the average shorter than those with entrances. Any comparisons between cave areas should rest on these as well as the known caves. Warwick (1968) has discussed from direct observations the importance of entranceless caves in a number of British karsts. For each karst Curl calculated a karst constant which secures the best fit between the observed and the predicted length/entrance data. No relation was found between karst constants and the accepted theories of origin for the caves in the different areas. But there was a good correlation between the constant and the mean length of all caves in each area, with the exception of one karst, that of Clare, Ireland, where the length frequency distribution was also anomalous.

Some of Curl's work has been criticised because so many assumptions lie behind the mathematical manipulation (Ollier 1963). There are indeed some quite unrealistic aspects to this work. For instance, if there is a water-filled passage in a cave, it is treated as two caves for the calculations. But the presence of the water is really a pointer to continued development of the cave as a whole. This objection does not apply to the same extent to treating barriers of rockfall or fluvial fill as the boundaries between separate caves since they may well evolve differently after such isolation. However, though these efforts may have been too broad and over-ambitious in a pioneer field, the basic method of applying stochastic models to assemblages of cave data, testing them and using them to direct further field inquiry is a valuable supplement to classic methods of karst study.

Lesser aspects of cave morphology may be more susceptible of mathematical analysis. No doubt stimulated by recent work on the meanders of surface streams, Ongley (1968) has analysed the directions of passage segments in Serpentine Cave, Jenolan, N.S.W. and shown that they are random. Preferred orientations which joint influence must have induced in the phreatic origins of this cave have thus been destroyed by subsequent solutional action. The tendency to oscillation is analysed by logging whether each segment of passage angles to the right or to the left of the previous segment. A series of right-bending segments followed by a series of left-bending segments constitutes an oscillation. Using 4-feet and 2-feet segments, the average wavelength was found to be 4·6 m, ranging from 1·8 to 7·6 m, and the wavelength/width ratio averaged 5·5, ranging between 2·9 and 8·5; the sinuosity was 1·4. He concludes there are no true meanders but that there is a definite meandering tendency.

UNDERGROUND SOLUTION

Whereas in the past chemical analysis of cave waters was chiefly used to help in water tracing (e.g. Oertli 1953), there has been in recent years increasing use of it to determine the course of solution underground. Water tracing has shown there is a tremendous range in speed of underground water movement. Gèze (1965) gives an average of 30 m/hour from French measurements. But the range is from 1000 m/h in steeply falling caves during storms to 4-5 m/h through low gradient caves with large storage in lakes and water-filled sections. Very variable speeds can be registered from the same system—6 to 500 m/h in the Sourciettes cave of the first type and 5 to 76 m/h in Padirac of the second type. So much more repetitive measurement than has been achieved so far will be necessary for a general understanding of cave solution.

However, several studies have shown that a great deal of solution by seepage water takes place close to the surface. Smith and Mead (1962) demonstrate this for a Mendip cave and Gams (1962) maintains most solution in Slovenia takes place in the top 10 m of limestone. Considering all available data from Clare, Ireland, Williams (1968) comes to the conclusion that as much as 80 per cent of solution by percolating water may occur within 8 m of the surface there with high concentrations of 200-300 mg/l. Pitty (1968) takes the same view for the Peak District with concentrations of the same order. In both of these British Isles karsts very shallow caves sometimes have much active deposition of dripstone confirming the measurements.

Whereas Smith and Mead (1962) found increasing amounts of limestone in solution with increasing depth of seepage water in G.B. Cave in the Mendips, Pitty (1966) found no correlation between depth and hardness in drips in Peak Cavern in the Peak District. A partial correlation analysis of the time series of observations from that cave showed greatest correlation between the amount in solution and surface temperatures 3-4 months before. Soil temperatures largely govern soil biochemical processes and the influence of the root zone, the 'rhizosphere effect', on CO_2 availability resulted in highest concentrations in the warmer summertime, together with a 'spring burst' of microbial activity and of solution. Although percolation waters in Postojna Cave in Slovenia had a different annual rhythm of carbonate concentration in the different climatic regime there, Gams (1966) stressed the important effect of surface vegetation in the matter. Seepage water was less rich in dissolved limestone under grassland than under forest but its concentration fluctuated much less. Two-thirds of the solution is attributed to biotic factors.

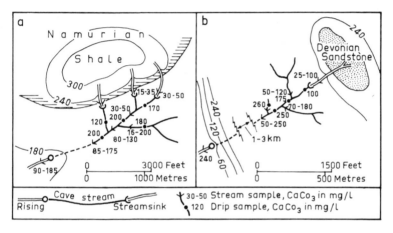

55 *Diagrams of solution in cave systems in*
   (a) Clare, Ireland.
   (b) Mendip, England. Both after Smith, High, and Nicholson 1969.

If so much solution takes place close to the surface, the question of how large caves develop at depth comes to the fore. Where there is inflow from impermeable rocks, vadose streamflow can account for this. Water analyses through Clare caves such as Poulnagallum (Smith and others 1969) suggest that the bulk of the water passing through these caves has come from higher shale terrain (Fig. 55a). On entry low carbonate values of 15 to 50 mg/l are measured. Along the main cave there are increments of dripwater and some tributaries are entirely of percolating water with concentrations averaging 170 mg/l. The result is that when little runoff from the shales enters the caves, the risings have high concentrations though still well below saturation, and when flood-waters enter the sinks the risings have concentrations only half as great. In Mendip, the surface catchments on sandstone and shale inliers are much smaller and contribute much less (Fig. 55b). Along G.B. Cave there is an increase of 20-100 mg/l along the main stream, the increase varying inversely with discharge; Smith and Mead (1962) think this is mainly due to drips and trickles and little to solution along the stream channel. As a result the risings here have a much more uniform concentration than the Clare ones. Smith and Mead (1962) postulated a steady increase of solute content along Mendip caves. However, Ford (1966) has shown that this is not so for Swildon's Hole in which a tributary contributes most of the increase. There is also considerable variation in carbonate content of the flows into the entrance of G.B. Cave, St Cuthbert's Swallet, and Swildon's Hole. Observations thus confirm theoretical expectations of much variation in the contributions of seepage water and streamflow of surface origin from cave to cave and karst to karst.

Because of practical difficulties of observation, the problem of solution at greater depth along water-filled passages (Bögli 1964a) remains one largely of theoretical discussion of mixing corrosion and possibilities for other renewals of aggressiveness high in the groundwater body.

VIII

CAVE DEPOSITS

Cave deposits of various kinds often occupy much bedrock cavity, affect cave form in other ways, and are also important because cave history is frequently better recorded by depositional evidence than by erosion, which is liable to destroy its own traces.

Genetic classification of cave deposits is difficult because there is much inheritance of characters from other environments and much mingling. Kyrle (1923) made a basic division between materials formed in place and transported sediments but this does not correspond simply with autochthonous and allochthonous origins (Kukla and Ložek 1958) since river deposits, for example, may be from either source. Others have gone straight to a division on the basis of mechanism of deposition, e.g. mass movement, water-laid, chemical and biological. Other mechanisms not so common include glacial, aeolian, and the freezing of water. Here space permits only an empiric and highly selective outline.

SPELEOTHEMS

Depositional forms growing from chemical precipitates are ambiguously referred to as 'cave formations'. 'Concretion' is too restrictive in its English connotation. 'Cave decoration' could be satisfactory but the American term, *speleothem,* is gaining general currency. The various processes of deposition are discussed in Chapter III; precipitation of calcite brought about by diffusion of CO_2 from water to cave air is the dominant one. However, the chief controls of growth of speleothems in Postojna Cave, Slovenia, are the amount and hardness of the seepage water, not variations in CO_2 content of the cave air (nor of evaporation, of course; Gams 1968).

Calcite deposited from drips before they are detached from
ceilings and walls forms downward growing *stalactites* and upward
growing *stalagmites* form where drips splash (Pl. 30). With straw
stalactites (Fig. 56a) the drip deposits a circular ring of calcite

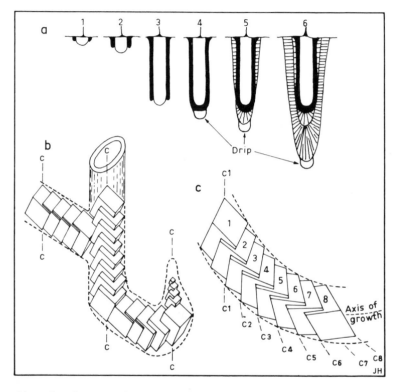

56 (a) *Development of a straw stalactite, blockage of capillary and growth
     to conical form.*
    (b) *Crystalline structure of angular helictite, with abrupt changes in
     growth direction. After Prinz 1908.*
    (c) *Curving helictite produced by gradual change in direction of c-axis.
     After Prinz 1908.*

to form a vertical tube about 5 mm in diameter. The greatest
recorded straw length is 6 m in Easter Cave, Augusta, southwestern Australia. Usually straws get blocked internally while
much shorter and then films of water flow down the outside to
build a conical stalactite with radially oriented crystals around
the central tube. Such stalactites may become very much longer

than straws; in Aven Armand, Lozère, France, one is over 30 m. The rate of growth varies in place and time in accordance with the drip rate, the state of the $CaCO_3$-CO_2-H_2O system, and the shape of the stalactite. A common rate of growth for straw stalactites is about 0·2 mm/year (Moore 1962). If the drip point shifts, a fresh stalactite usually forms but with stalagmites which are usually much broader in relation to height because of splash and flow, a shift in the drip point simply leads to a complex doming. Very slow drips may inhibit stalagmite growth, whereas too fast a rate may prevent stalactite formation.

A distinction is sometimes made between dripstone and flow-stone. However, all but the simplest stalactites involve flowing films as well as drips. When a stalactite and a stalagmite meet to form a column (Pl. 30), dripping is eliminated but not growth. Curtains and shawls are due chiefly to trickles down walls. Rim-stone dams or gours grow up on floors to impound pools of various sizes and depths. Aeration over a growing rim promotes loss of CO_2 and calcite precipitation. Siffre (1959) has invoked laminar flow as an essential condition for gour formation. They form in rivers as well as from thin flows over floors as is well exemplified over a kilometre of stream in Croesus Cave, Mole Creek, Tasmania, but deposition may only occur when flow is small. Calcite deposits in rimstone pools often have an external crystalline form reflecting their internal structure such as dog tooth spar. Floe calcite forms as a thin film on pool surfaces and can accumulate on the bottom.

Helictites are eccentric speleothems which defy the law of gravity, growing in any direction from solutions fed along tiny capillaries. Angular helictites involve abrupt changes in crystal lattice orientation (Fig. 56b), but gradual changes in axis of curving forms are as common (Fig. 56c; Prinz 1908). Some helictites consist of or are covered in calcite crystals with external crystalline form without ever having been immersed in standing water. Many forces and processes have been cited to explain helictites (Moore 1964) but it cannot be said that any explanation has yet acquired general support. Curving crystal forms are very common in gypsum and halite speleothems (Lowry 1967b). Many other kinds of speleothem occur and are discussed in the literature (Warwick 1962).

Speleothems occur in actively developing river caves but their formation can become dominant in passages and chambers abandoned by streams. It has been claimed that speleothems form rarely when water temperatures approach freezing (Corbel 1960), that stalactites are more frequent relative to stalagmites in cool temperate climates (Corbel 1952), and that stalagmites and bulkier forms generally assume dominance with increasing warmth in tropical caves (Corbel 1959b).

CAVE ICE

In snow climates, snow drifts into potholes. Thick banks may block shafts and stop warm air circulation in summer. Parts of the drift may survive summer melting and turn into ice masses of semi-permanent nature, often conical in form.

More diversified ice bodies accumulate in the lower parts of caves where cold air enters in winter and freezes descending seepage water, e.g. in Eisriesenwelt and Rieseneishöhle near Salzburg, Austria. Floor ice may be as flat as an ice rink or steeply but smoothly sloping. It is misleading to label such ice masses underground glaciers because they are immobile; the internationally used French term *glacière* designates caves with ice in them (Balch 1900). Icicles, stalagmites, and columns of ice are broadly equivalent in form to their speleothem counterparts and may become very big. They can form and degenerate over comparatively few years (Kyrle 1923).

BIOGENIC DEPOSITS

Animals living completely in caves are small in size and even as large populations cannot build significant accumulations, so biogenic deposits are ultimately derived from outside caves. Large masses of excrement are piled up by colonies of larger animals which feed outside caves. Bat and bird guano in great conical masses beneath roosts in chambers are the chief materials of this kind.

Animal bones also accumulate. They may be from animals which sheltered in caves (e.g. in Mixnitz Cave, Austria, 500,000 individual cave bear skeletons have been counted) or from animals that fall accidentally into potholes and vertical shafts.

The prey of carnivorous animals also contributes to cave bone deposits, including the pellets of owls and food refuse of human kitchens.

With time these organic materials are mineralised. Phosphates, nitrates (saltpetre), and other minerals are formed.

Primary biogenic accumulations are subject to water transport at any stage and may be incorporated in clastic sediments. In most well populated countries, biogenic deposits have usually been mined for fertiliser and, in the case of saltpetre, for gunpowder, so that the floors of caves from the United States through Europe and Asia to Australasia which had such deposits are now in a very modified state through removal and translocation.

CLASTIC SEDIMENTS

Clastic sediments vary a great deal in caves because of diverse origins, modes of transport, and environments of deposition.

Breakdown deposits. Breakdown of roofs and walls usually results in massive crudely conical heaps of angular blocks of rock on cave floors (Pl. 3; Pl. 38), though some blocks may be partly rounded by solution before detachment. Finer grained aggregates result from salt weathering of the surfaces of dry caves. Textural sorting is generally poor, and textural parameters are frequently bimodal because inherited textures are superimposed on breakdown textures.

Gash-breccias and bone-breccias are exceedingly characteristic of caves and have long occasioned interest (e.g. Pengelly 1864). These are very mixed deposits with large bedrock fragments in clay matrix and sometimes with sand. Bones and also speleothem materials, which may be in place or transported, may also contribute. These breccias come from karst bedrock, insoluble residues, surface soil materials, secondary precipitates, and animal remains sliding under gravity down fissures and shafts. They are found mainly as fissure fills but also in cave chambers, in part because of cave enlargement since their emplacement, as in Dip Cave, Wee Jasper, N.S.W. (Jennings 1963).

Stream deposits are commonly of external origin but insoluble residues, rockfall, and speleothem fragments may be shifted around within caves by streams. Well sorted and rounded sediments of greatly varying modal size from boulders to clay result. The size frequency is normally skewed to the fines because of clays inherited from the insoluble fraction of the bedrock. Mineral composition is very varied where the cave streams are bringing in materials from outside the karst. Calcite sand is rare but dolomite commonly yields sand-sized sediment, e.g. in Grotte de Moulis, Ariège, France. Quartzose sands may be from karst bedrock or from surrounding impervious rocks.

Many sedimentary structures are encountered, such as cross-bedding and cut-and-fill. Graded beds, the products of big floods into the cave, are common, not only in areas which suffered glaciation, with attendant glacial meltwaters, but in a wide variety of karst environments. Frank (1971) describes graded beds of 19,000 B.P. from Koonalda Cave in the Nullarbor Plain; mineralogically these do not differ from similar sediments being deposited now in the present semiarid climate so he finds no evidence in this for a former wetter climate. Lamination may be especially common in cave stream sediments because of fine fractions supplied by insoluble residues and of great fluctuations in flow (R. M. Frank, pers. comm.). Siffre and Siffre (1961) have presented shape data of limestone pebbles which have passed through inverted siphons and argue that pressure flow results in highly rounded but flattened forms. Cementation of cave stream deposits is common because of frequent occurrence of super-saturated carbonate waters.

As well as active channel deposits, abandoned stream deposits are frequently found in caves, e.g. as terraces. Some caves may be practically filled with gravel like Baldocks Cave, Mole Creek, Tasmania, where the fill is probably of Pleistocene glaciofluvial origin. However, too frequently inference of thick or complete fill has been made from small patches of deposit in high positions in cave walls. Often these are no more than slip-off slope deposits and meander niche fills which have no such implications, but register only former levels in cave incision and stream lowering (Fig. 53j). Floods can deposit fine sediments on ledges at many levels at the same time. Alluvial passage blockages sometimes deflect cave streams along different courses or cause a fresh channel to be cut in bedrock above the fill (Fig. 53f).

Red clay. Red or ochreous clays, sometimes unctuous, are found widely in limestone caves in many climates; within cave systems they are probably most frequent in phreatic sections. They may include silt and fine sand but clay minerals predominate, kaolinite being the most common. They are unbedded but liable to desiccation cracks, into which other materials may be introduced subsequently. The reddish colour is usually attributed to ferric iron.

In the past these clays were regarded as a product of internal weathering of cave bedrock surfaces. But Bretz (1942) thought that in many caves the volume of limestone removed could not have provided the necessary bulk and he maintained they were derived from surface soils and were deposited in stagnant phreatic conditions. Detailed work by Reams (1968) on the Ozark Mountains caves shows that the caves are shallow phreatic in origin but that the sediments were deposited by streams bringing in surface materials after the surface rivers had incised their valleys and a vadose phase had supervened.

In Austria and Switzerland, cave loam (*Höhlenlehm*) was similarly regarded as cave weathering residue. The term has been applied to a range of sediments from sandy loams to heavy clay. Bögli (1961b) has shown that cave loam in the Hölloch system, Muotatal, Switzerland, was brought in from the surface by streams and only calcium carbonate cement is spelean in origin.

ENTRANCE FACIES

Deposits in and near cave entrances have been studied more than all other cave deposits, partly because of their accessibility but more because of their valuable complex stratigraphic sequences and archaeological associations (Schmid 1958).

Here surface materials are moved in by various mechanisms and interbedded with materials originating in the cave. Surface soil, weathering mantle and bedrock may fall, slide and creep, especially in periglacial conditions when regolith is exceedingly mobile and frost wedging provides much angular rock. Pleistocene glaciers have forced moraine into cave entrances and potholes; these often form a substratum for later accumulations. Varves may be deposited by glacial meltwaters farther in. Entrance facies may receive aeolian accessions in the form of loess near Trade Wind deserts and near glacial outwash plains in higher latitudes in Pleistocene cold periods. Bones and excrement register phases of occupation of cave entrances. Fossil assemblages may reflect external environmental changes in some detail; mollusca have been particularly useful in this way (Ložek and others 1956). Men have similarly left a physical record of their occupation behind, with tools and hearth ashes as well.

Roof fall may be intercalated between these layers of allochthonous material or be incorporated in them. Periglacial phases may accentuate rockfall because frost wedging can reach tens of metres into caves in rigorous climates. In addition, calcite floors may be precipitated over entrance areas and authigenic minerals are found in the voids of earlier sediments, chiefly calcite, aragonite, gypsum, salt, and phosphates.

Bedding can be complex in the entrance facies, frequently with high depositional dips. Nevertheless the variety of deposits laid down in sequence has provided much evidence for Quaternary history in many countries.

Riedl (1961) has stressed the importance of extending the manifold field and laboratory analyses applied to the entrance facies to all cave deposits and of relating these as closely as possible to speleogens on all scales. In this he stresses the importance of a climato-morphogenic approach. Isotopic studies of speleothems are being intensified with promising results. Hendy (1969) has combined C-14 age and oxygen palaeotemperature determinations of stalagmite calcite to produce a palaeotemperature curve over the last 100,000 years for the Waitomo caves area of New Zealand, whilst Duplessy and others (1970) have constructed a similar record for the Riss-Würm interglacial (130,000-90,000 B.P.) from a stalagmite 3 km inside Aven d'Orgnac, France.

IX

INFLUENCE OF CLIMATE AND RATES OF DENUDATION

Classical ideas about karst arose chiefly from studies of the Dinaric karst and that of central Europe. Differences between them were appreciated. For instance, central Europe was noted for possessing a greater proportion of covered karst than the Dinaric region. In order to explain these variations, climatic factors such as frost incidence and rainfall intensity were invoked (Grund 1910b), but the possible relations were obscured by complications due to forest clearance and geological structure. The dominant mode of thought as exemplified by Sawicki (1909) and Grund (1914) was to range all karst phenomena into a single evolutionary sequence or cycle. Even strikingly different karst forms described from tropical humid Jamaica and Java (Daneš 1908, 1910) were regarded as falling into place in such schemes, and landform terms from the Dinaric karst were applied without much discrimination the world over, almost with the effect of mental blinkers (Lehmann 1960). With the emergence of general concepts of climatic morphology in the 1930s came a clearer realisation that karsts in contrasting climates might have independent modes of development. This was most evident in a study of Javanese karst by H. Lehmann (1936) but had been foreshadowed in the work of O. Lehmann (1927) on changes of karst features with altitudinal climatic zonation in the Austrian Alps.

RATES OF KARST DENUDATION

After World War II geographers interested in karst became pre-occupied with climatically controlled morphogenic systems; at the 1953 meeting of the International Geographical Union's Karst

Commission (Lehmann and others 1954) no fewer than eight climato-morphogenic regimes were discussed. Corbel in particular (e.g. 1957) pursued this question and he gave it a quantitative slant by advocating measurements of rates of karst denudation (Corbel 1959a). His formula

$$\text{Limestone denudation (m}^3/\text{km}^2/\text{year, or mm}/1000 \text{ years)} = \frac{4 \, E \, T \, n}{100} \text{ where}$$

$E =$ runoff in decimetres,

$T =$ mean $CaCO_3$ content in mg/l

$\dfrac{1}{n} =$ fraction of catchment in limestone and limestone alluvium

has subsequently been improved by Williams (1963) and Douglas (1964) by allowing for departures from assumed density of 2·5 for the karst rock and by substituting total carbonate hardness for calcium carbonate alone. Greater difficulties arise in obtaining adequate field data. Corbel frequently employed far too few figures of solute content for the mean to be representative and, since discharge data were frequently unavailable, annual surplus of the water balance (precipitation minus evaporation) was substituted. Climatic figures for evaporation are notoriously problematic and abnormal infiltration in karst affects this calculation particularly. Table 6 presents a selection of karst denudation rates from different climates, about half of which are based on discharge values and sufficient chemical determinations.

In recent years there has been some reaction to what is claimed as overemphasis on karst differences attributed to climate (Panoš and Štelcl 1968), moreover detailed studies reveal sharp variation in spatial and temporal distribution of solution within given karst situations (Gams 1966; Douglas 1968; Pitty 1968; Smith and others 1969). Undoubtedly there has been a tendency to overlook greater differences due to other causes such as lithology in the search for the operation of the climatic factor and to claim greater interpretative certainty in the face of the complex interactions with many other factors. Nevertheless the broad reality of climatic control in karst cannot be denied, even if the nuances have not been evaluated.

TABLE 6 **Rates of karst denudation**

Karst area	Köppen climatic type	Mean annual precipitation (mm)	Net rate of denudation $m^3/km^2/y$ or $mm/1000\ y$	Source
Somerset Island, N. Canada	ET	130	2	Smith 1969
Tanana R., Central Alaska	Dfc	450	40	Corbel 1959a
Svartisen, N. Norway	Dfc	740–4000	275–5000	Corbel 1957
Vercors, France	Dfb	1500–2500	240	,, ,,
Punkva R., Moravia, Czechoslovakia	Dfb	620	25	Štelcl and others 1969
Fergus R. and Shannon R., Ireland	Cfb	1000–1250	51–53	Williams 1963, 1970
Craven, England	Cfb	1250–1500	40	Sweeting 1965
Peak District, England	Cfb	800–1200	75–83	Pitty 1968
Mellte R., Wales	Cfb	1600	16	Groom and Williams 1965
Mendip, England	Cfb	900–1100	40	Corbel 1959a
Slovenia, Yugoslavia	Cfb	1250–2000	10–100	Gams 1966
S. Algeria	Bwh	60	6	Corbel 1957
Los Alamos, New Mexico, U.S.A.	Bwh	25–40	< 1	,, ,,
Grand Canyon, Colorado, U.S.A.	Bwh	25–50	7	,, ,,
Kissimee R., Florida, U.S.A.	Cfa	1200	5	Corbel 1959a
Yucatan, Mexico	Aw	1000–1500	12–44	,, ,,
Indonesia	Afi	200–3000	83	Balázs 1968

THE ARID EXTREME

Water is necessary for karst development; vegetation and soil microbial activity promote it, chiefly by providing much more CO_2 than the atmosphere alone. Regions of low precipitation, particularly those of high temperature where evaporation renders

31 *Kestrel Cavern No. 1, Nullarbor Plain, Western Australia. Extensive, almost featureless surface of retarded arid karst, with rare collapse dolines which retain freshness of form for long periods.*

32 *Hemispheroid conekarst in Miocene limestone at Iehi, Lower River Kikori, New Guinea. Partly cleared of rainforest.*

rainfall less effective, have poorly developed karsts in consequence.

The Nullarbor Plain in Australia, one of the world's largest karsts, with not far short of 200,000 km^2, is entirely semiarid and arid in climate and much of it is treeless as its name implies. Uplifted in Miocene times with only gentle tilting, it remains an unbroken low plateau (Pl. 31). There are virtually no valleys but a very gentle shallow undulatory relief, generally no more than 3-6 m in amplitude, is a product of differential surface solution, guided by joints. There are a few collapse dolines which long retain their sharpness. Minor surface sculpture is minimal and best developed near the coast where sea spray drifts on to the rocks. Less than a score of large deep caves are known which reach down to a flat watertable with brackish slowly moving water in a lower highly porous chalky limestone. Shallower caves are more numerous but the underground as well as the surface karst is greatly retarded despite ample time since emergence (Jennings 1967c). Gypsum and halite speleothems assume a greater than normal importance as does salt weathering in caves (Lowry 1967b). There is a great deal of calcrete induration of surface limestone.

These characteristics are common for karst in hot, dry climate (Wissmann 1957; Conrad and others 1968). Corbel (1959) has reported low rates of present denudation in accordance with these traits (Table 6). Where there are regional relief contrasts of tectonic origin or inherited from previous geomorphic phases, limestone or dolomite karst acts as resistant rock because it is almost immune from surface water erosion. Evaporite karst persists at the surface more than in humid climates and even salt domes provide projecting relief for some time after stripping, then leading to craterlike forms and salt swamps (Harrison 1930).

THE COLD EXTREME

In cold climates where water on the surface and in soil and rock is frozen most or all of the year in association with geophysically polar glaciers and with permafrost, karst development is also inhibited. Beneath glaciers, claims Corbel (1954), limestone acts as resistant rock but resultant landforms comprise glacial, not karst relief. In the permafrost zone, underground karst development stops and even though there is surface water in a short summer, total amounts are low. Frost wedging destroys incipient surface solution features and shillow clutters outcrops. Because structural forms are prevalent, it is thought that the surface is lowered parallel to itself, though solution is more active beneath snow banks than along streams. According to Corbel, surface lowering proceeds slowly because of low available amounts of water and, according to Smith (1969) on the basis of observations from Somerset Island in northern Canada, because of poverty in biogenic CO_2 through low vegetative productivity. However, Corbel (1959) reports extremely high carbonate contents from deep wells below permafrost levels in northern Alaska where karst processes may be slow but have not stopped.

Conditions are different in maritime and high mountain cold climate and in cool temperate snow climate. Here glaciers are accompanied by abundant meltwater and bedrock temperatures permit infiltration of water. Beyond the glacier margins, frozen ground is only seasonal and snow melt provides seasonal abundance of liquid H_2O. In these conditions there can be vigorous development of karst. Close-set fields of small solution dolines and potholes occur beneath forest and in alpine meadow and

33 Close set potholes and dolines in Palaeozoic marble, Mt Arthur, Nelson, New Zealand. Shillow on outcrops due to frost shattering.

tundra conditions (Pl. 33) though they may be subordinate to larger glacial forms. In the Queyras district of the French Alps, glacial rock basins in dolomitic limestone lack lakes through underground drainage and their glacial moraine cover is pitted with subsidence dolines. Minor surface solution sculpture is sometimes well developed, but the effects of recent ice action may not have been removed or frost shattering may still be too persistent for the development of minor forms. Some kinds of *Karren*, e.g. solution funnel steps, appear to be genetically linked to snowmelt (but they are also found on the Irish coast (P. W. Williams 1966a)).

Cave development can be extensive and in high mountain karsts of this type, very deep systems have developed, especially in the Alps and the Pyrenees. The caves are usually poor in speleothems and glacières occur. Corbel (1957) claims that in these climates limestone acts as a weakly resistant rock, though young tectonism may prevent it from finding present expression. In northern Norway and western Tasmania, limestone outcrops are disposed along valley floors and karst has largely been destroyed in these older tectonic milieus.

There is argument about the cause of advanced karst in these kinds of climate. Corbel (1954, 1957) has stressed higher CO_2 saturation equilibria of cold waters and the presumed high CO_2 content of the voids in snow and ice. However, a more common view is that saturation values rarely control development and the kinetics of solution are more important. Observations by Ek (1964) on ice melt and meltwater rivers in the French mountains show higher pH and lower carbonate concentration than Corbel claims; they do not testify to special aggressiveness. High precipitation and low evaporation are admitted by all as explaining high denudation rates in karsts such as those of Svartisen in Norway and Vercors in the French Alps. Even in the inland taiga-tundra basin of the River Tanana in Alaska, with a modest annual precipitation of 450 mm, the bulk of solution takes place during the short summer season of great floods when carbonate concentrations are lowest (Corbel 1959a). Sheer volume of water and its turbulence dominate solution.

THE 'BOTANIC HOTHOUSE' EXTREME: TROPICAL HUMID KARST

The most important fact about karst in tropical humid climates with or without a dry season is its greater variety than elsewhere

(Verstappen 1960a, 1964; Jennings and Bik 1962). The many type names applied—cockpit karst, tower karst, labyrinth karst, crevice karst, polygonal karst, arete and pinnacle karst are examples—testify to this. That climatic factors are not the only ones involved is illustrated by the range of explanations given for doline karst in hot, humid climates. These consist of dolines perforating planation, structural or tectonic surfaces in a manner and to a degree similar to that found in classical mid-latitude karsts. Such doline karsts have been attributed in the tropical humid context to marly limestone or to interbedding with impervious rocks as in the Aguada Limestone of Puerto Rico (Monroe 1968); to mechanically weak limestones as in Yucatan; to longer dry season as in northern Jamaica; or to closeness to sea level as in subtropical Florida.

Great variety is a pointer to extremely active karst development and many investigators have interpreted the most striking kinds of tropical karst as the product of rapid, vigorous solution. These are the kinds now commonly subsumed under *conekarst* (Ger. *Kegelkarst*) in which the relief is dominated, not by closed depressions, i.e. reduction forms, but by projecting residual forms, rarely, however, of conical form (Fig. 57, Pl. 32). Two varieties have been contrasted in the literature, though in reality they grade into one another in some contexts and there are many areas which conform to neither (Aub, in press a).

Cockpit karst is typified by the Gunung Sewu karst of south central Java (Lehmann 1936; Flathe and Pfeffer 1965), though the name comes from similar country in Jamaica (Lehmann 1953; Sweeting 1958). Here the residual limestone hills are characteristically hemispheroidal in shape *(Kugelkarst)*. The Gunung Sewu are about 30 to the square kilometre and 30-70 m high. The closed depressions between them, the cockpits of Jamaica, are often of star shape, with sides bulging inwards (see p. 134; Fig. 43). Often the hills are aligned and the cockpits strung out in chains along 'glades' between. The relief is clothed naturally in dense rainforest; red earth and rendzina soil (Pfeffer 1969) cover variable proportions, sometimes confined to the depression floors, whilst forest litter patchily hides craggy slopes. Despite the forest, water may course down the slopes during intense rainfall of tropical storms, especially where adjacent slopes converge.

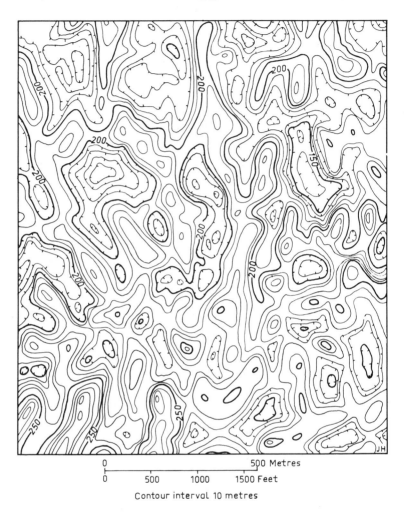

0 ⊢———⊣ 500 Metres
0   500   1000   1500 Feet
Contour interval 10 metres

57 Conekarst in northern Puerto Rico. From U.S. Geol. Surv. map.

Lehmann (1936) interpreted the Gunung Sewu cockpit karst as due to rejuvenation of surface drainage by tectonic updoming of a planation surface on limestone (Fig. 58). This produced ridges and valleys aligned down the domes. Then with the development of vertical infiltration of the water and underground drainage, the valley systems were broken up into chains of closed depressions of surface solutional origin. Detailed work by Aub (in press b) in a small area of Jamaican karst confirms a surface solutional origin of its cockpits. In other cockpit karsts, alignment of hills and cockpits is controlled by jointing, not necessarily related to the latest tectonic uplift, and in others, e.g. along the lower Kikori River, Papua, the former presence of a planation surface is in doubt. These differences, however, do not preclude a basically similar evolution.

Tower karst (Ger. *Turmkarst*) differs from cockpit karst in much steeper lower slopes to its residual hills, which can be vertical or overhanging, and in the presence of swampy alluvial plains around the towers (Fig. 59, Pl. 34). The towers themselves may surround flat-floored depressions, basically of polje type. Karst margin plains frequently accompany tower karst and deep karst corridors, vertically walled linear depressions, often cut up

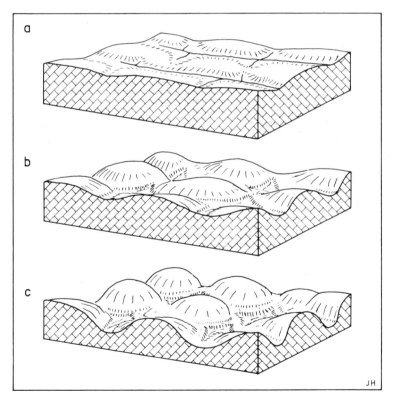

58 *Evolution of cockpit karst of Gunung Sewu type according to Lehmann 1936*

the hill masses. Surface drainage over the flat plains may predominate areally. Horizontal tunnel caves through the towers link
the poljes with plains which have surface drainage to the sea and
the levels of their outflows *(Vorfluters)* regulate the polje floor

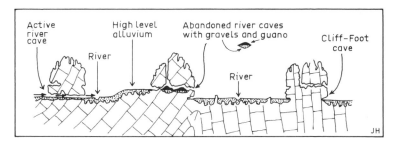

59 Section through tower karst based on Kinta valley, West Malaysia

*34 Tower karst in Permo-Carboniferous limestone in Perlis, West Malaysia.
Towers rise 360 m above alluvial plain.*

heights. Inactive higher level caves are the sites of guano accumulations and phosphates (Wilford 1964).

Most Malayan karst, e.g. in the tin and iron mining Kinta
valley with towers up to 500 m high, is of this type and so also

that of Sarawak (Wilford and Wall 1965). Tower karst is described from Cuba where the towers are called 'mogotes' (Lehmann and others 1956) and southwest China and Vietnam (Wissman 1954; Gellert 1962; Silar 1965).

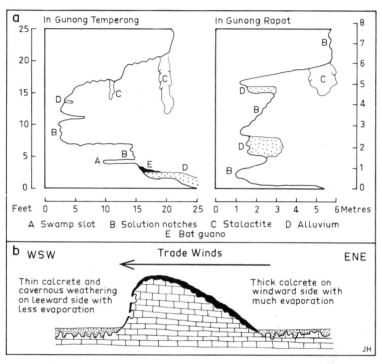

60 (a) *Cliff-foot cave sections from Kinta valley. After Walker.*
   (b) *Mogote asymmetry in Puerto Rico resulting from differential calcrete formation according to Monroe 1966.*

Tower karst is generally attributed to lateral solutional undercutting by river floods and swamp waters around the margins of alluvial plains, which tin, iron and gold mining in Malaya and Sarawak, for example, have shown to overlie limestone surfaces, intricately cut up in detail by subsurface solution but planed off flat overall. Cliff-foot caves (Fig. 60a) are frequent round the bases of towers though by no means present everywhere; these are low-roofed, narrow caves elongated along the tower margins with occasional passages leading more deeply inwards. Swamp slots and solution notches (Chapter IV) are also common, frequently in association with depressions marginal to the alluvial plains. Many tower karsts give strong expression to the tendency for karst solution to operate dominantly in vertical and horizontal planes.

Whether it is cockpit or tower karst which develops in any area has been variously explained. In Tabasco, Mexico, tower karst is thought to follow cockpit karst as a later stage of development (Gerstenhauer 1960) whereas in southwest Celebes nearness to impervious basement is the control preferred, shallow karst favouring cockpit karst (Sunartadirdja and Lehmann 1960). Renault (1959) thinks that less compacted, more jointed and faulted rocks support cockpit karst only in Kouilou, Gabon, but for West Irian, Verstappen (1960a) relies on variation in porosity as the determinant, the more porous giving rise to cockpit karst. In Puerto Rico, presence or absence of a former cover of superficial deposits is regarded as critical, cover resulting in tower karst (Monroe 1968).

In Puerto Rico there is asymmetry in the residual hills that is not related to attitude of the limestone but to the prevailing winds; Thorp (1934) attributed it to greater rainfall and consequent solution on the windward eastern side, producing gentler slopes there. According to Monroe (1966), however, the mechanism is one of greater superficial induration of weak limestone on the windward side (Fig. 60b). More wetting and drying takes place there with greater solution and reprecipitation. The thinner caprock of the western sides allows readier attack and undercutting of the weak interior to steepen these flanks. His evidence makes plain the importance of case-hardening of this type in the resistance of the residual hills to denudation here but many other tower karsts, e.g. most Malayan karsts in dense Palaeozoic limestones, owe little or nothing to such processes.

Tropical limestone towers are not always set in alluvial plains. In Malaya many stand with customary abruptness on steep granite slopes (Fig. 62) and in New Guinea (Jennings and Bik 1962) they often rise from supporting conical or pyramidal limestone hills. How the cliff-foot angle is initiated and maintained is easier to understand in the first case (because water must issue at the limestone-granite contact) than in the second circumstance. Nor are the plains associated with towers always developed on karst rocks. For example, tower karst surrounds some of the 'interior valleys' of Jamaica, large closed depressions of which the floors are entirely inliers of an impervious basement.

These variations do not invalidate the general view that enormous amounts of limestone have been removed rapidly to create these elaborate karst styles. Most authorities have explained this in terms of the climate and the vegetation (Lehmann 1953; Birot 1949). Chemical reactions proceed faster because of higher temperatures; high rainfall and high intensities of rainfall make for prolonged and rapid solution; rapid plant growth and decay and intense microbial activity with high soil Pco_2 make tropical water very aggressive.

Corbel (1959a) opposed this 'botanic hothouse' interpretation on the grounds of lower saturation equilibria for CO_2 and carbonate solution with high temperatures and of low observed carbonate concentrations in tropical karst waters. He explains the undoubted elaboration of these karsts by the great length of geological time which has been available for uninterrupted development, whereas Pleistocene glaciations have intervened in many high and mid-latitude karsts, some of which have developed entirely in the Holocene. Such histories may be true in certain cases; in Clare, Ireland, the caves and underground drainage are of Postglacial age (Ollier and Tratman 1969); Malayan karst probably reaches back into the Tertiary in continuous development. On the other hand some mid-latitude karsts, e.g. the Dinaric, have inherited some features from the Tertiary and there are conekarsts in New Guinea which were not uplifted till the Pleistocene and are very young (Jennings and Bik 1962). There is much reprecipitation of carbonate in humid tropical karsts so that net removal rates may conceal greater absolute solution. But it is probable that saturation equilibria rarely control action here. More detailed studies of discharge and water chemistry in high precipitation tropical karsts are needed.

Nevertheless the complexities of tropical humid karst will not be solved in climatic terms alone as is evident within one fairly uniform tropical humid climate with a short dry season in Perlis and Kedah states of Malaya. Along strike belts of Permo-Carboniferous limestone there are lines of great towers rising up to 360 m high from coastal alluvial plains (Pl. 34). Immediately west of this typical tower karst there is an anticlinal belt of Ordovician-Lower

Silurian limestone along the Thailand frontier; much of it is of high purity though some is dolomitic and some impure. In the main it forms a massive plateau area of 300-600 m, which is predominantly a doline and polje karst. The poljes are deep and steeper-sided than typical Yugoslavian ones but otherwise much resemble them. Small dolines pepper the plateau surface. However, a lithological explanation is not immediately apparent because the plateau karst is continued along identical strike belts into the Langkawi Islands across a very shallow sea (Pl. 39). Some of this karst is comparable but there is tower karst also, some rising from mangrove swamps and some from the sea. Though marine notches indicate present-day recession by seawater solution of the latter, it is unlikely that the presence of tower karst can be the result of the same conditions as are now found. The presence of large, round dolines in the island of Pulau Dayang Bunting which are occupied by freshwater lakes, suggests that the karst may have formed in part during lower sea level stands. Only combinations of factors explain this complex of karst types in northwestern Malaya.

35 *Vertical walls and aretes round dolines in Miocene limestone on Mt Kaijende, Central Highlands, New Guinea*

FURTHER KARST CLIMATO-MORPHOGENIC SYSTEMS

Corbel (1957) has sought to distinguish a karst style in cool temperate maritime climates as in Ireland where Williams (1970) has pointed to the importance of peaty waters rich in organic acids and where there are persistent light rains and low frost incidence. However, Corbel cites few distinctive forms other than badland sculpture of scarps exposed to wind-driven rains but not subject to frost wedging. Tropical subhumid and semiarid karsts in Brazil (Tricart and Silva 1960) and in northern Australia (Jennings and Sweeting 1963) present some traits of tropical humid karsts, e.g. evolution to tower karst, but others of drier climate affiliations, including pedimentation, are also present. Whether this can be explained entirely on the basis of present climate is not certain (Jennings 1969). Indeed all efforts to distinguish climato-morphogenic systems other than those associated with climatic extremes must be regarded as tentative in terms of present knowledge (Pl. 35).

X

INFLUENCE OF GEOLOGICAL STRUCTURE

Geological structure is so powerful and pervasive an influence on karst that none of the preceding chapters has escaped reference to it. Nevertheless some separate treatment of it is necessary. Here structure is taken to include both original depositional structure of the rocks with their varying lithology and later tectonic features such as folds, faults, and joints.

Primary depositional and tectonic forms are more evident in karst than with many other rocks, and karst regions appear disproportionately in landform atlases and geomorphological texts to illustrate tectonic relief. This has often been explained as due to a special immunity from surface attack conferred by the development of underground drainage. Though such karst immunity may be true in arid and semiarid climates, many measures of surface solution of limestone in other climates are of magnitudes which preclude long persistence of surfaces intact. Thus Aubert (1969) has argued that if the present low rate of surface ablation of 0 05 mm per year in the Jura Mountains has been maintained since the Pontian orogenic movements, 500 m of limestone have been removed. Lowering of surfaces along with preservation of form by uniform reduction by solution may resolve this paradox in some cases (P. W. Williams, pers. comm.).

DEPOSITIONAL RELIEF IN REEF AND DUNE LIMESTONE KARSTS

Biogenic reefs are particularly illustrative of this attribute. Verstappen (1960b) describes various types of primary relief on emerged 'coral' reefs, according to steepness of the basement on which reefcaps rest, vigour of reef growth, and course of uplift, and he also discusses their destruction by solution (Fig. 61). Rapid emergence of fringing reefs on the north Huon coast of New Guinea has left a sequence of marine constructional terraces of well preserved form in a humid tropical climate rising to over 500 m and ranging back to 450,000 ± 100,000 B.P. (dated by $U^{238/234}$ (Veeh and Chappell 1970)). Narrow, close-set sequences relate to times of rapid negative movement when tectonic uplift and falling eustatic sea level worked in opposition; broad terraces with lagoon facies as well as reef rim mark stillstands when rising land and rising sea levels were matched.

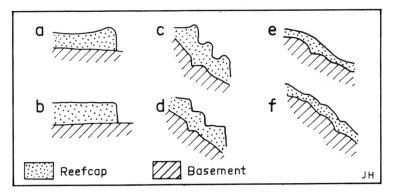

61 *Types of emerged reef relief according to Verstappen 1960a*
  (a) *Plateau reefcap on broad platform with good reef growth.*
  (b) *As above but with poor reef growth.*
  (c) *Terraced reefcap on steep basement with intermittent uplift and good reef growth.*
  (d) *As above but with poor reef growth.*
  (e) *Undulating reefcap with thin reef and no reef edges due to rapid uplift.*
  (f) *As above but with occasional reef edges.*

Emerged atolls such as Rennell Island in the Solomons and the Trobriand Islands (Ollier and Holdsworth 1968, 1969, 1970) have an annular outer ridge which was the former reef and a swampy interior plain which was the former lagoon. The coastal cliffs of the rim have certainly suffered sea attack, witnessed by marine notches at various levels, but they also owe much to the primary precipitous nature of the outer face of the reef.

Active reef growth tends to incorporate many primary voids, which may be quite large, and there is the possibility that emerged reefs have inherited caves from marine conditions. However, Ollier and Holdsworth found no evidence of this in the Trobriands where caves commonly are of single level development by solution close to the present watertable, followed by collapse and much speleothem development, though there is surprising variety. On some emerged reef islands caves develop as tunnels leading radially to the sea such as Tumwalau Cave on Kiriwina and Bwabwatu on Kaileuna. Most are much modified by collapse. On Kitava several large caves have large symmetrical scallops and bellholes indicative of phreatic development; a freshwater lens may have reached below the sea level of the time of their formation.

In Bermuda in a limestone formation part reef, part aeolian, there are caves reaching below sea level with submerged speleothems: according to Swinnerton (1929) they formed during a period of lower sea level in the Pleistocene. Bretz (1960) maintained that when the island was larger with lower sea level, it was big enough to maintain a body of fresh groundwater that developed the caves phreatically. Further lowering allowed air to enter which permitted speleothem growth. But on the small Trobriand Island of Vakuta the caves seem to have developed in brackish water so that the establishment of a fresh groundwater body does not seem to be essential.

The intricate pattern of sounds and bays on Bermuda is entirely attributed to cave solution and collapse in Pleistocene glacial low sea levels by Bretz, but Swinnerton thinks only some like Castle Harbour are of this origin, others being simply due to the flooding of primary hollows in dunes deposited during the low stillstands.

It is evident from much of the western and southern coasts of Australia that primary dune forms can persist through the consolidation of calcareous dune sands to aeolian calcarenite (Jennings 1968). There are multiple sequences of these dune systems and those on Rottnest Island off Perth, Western Australia, dated at 100,000 B.P. and retaining strong primary relief, are not the oldest. For Bermuda, both Swinnerton and Bretz postulated that diagenesis of aeolian sands preceded karst development. However, in the Australian dune limestones development of some karst features has accompanied consolidation (Jennings 1968). Some caves occur in only partially consolidated sands even today. This partly syngenetic karst has some distinctive features: calcite deposition contributes to solution pipe development (p. 50); some gorges are depositional in origin (Pl. 36); collapse plays a precocious role in cave development; the form of the crystalline basement beneath the dunes affects cave formation also as does differential lithification.

STRUCTURAL RELIEF IN REEF KARST

Even in ancient reefs, structure may remain evident through its guidance of solution as is well illustrated by the Devonian reefs of West Kimberley (Fig. 3b). Only modestly disturbed by tectonics, these were buried beneath Permian sandstones and then were truncated by a Mesozoic-Tertiary planation surface. Subsequent uplift and rejuvenation have led to removal of the weaker impervious basin and inter-reef facies so that the barrier and patch reefs stand out now from shale plains as the low plateaux and ridges of the Limestone Ranges. Though pedimentation has cut into reef structures to margin the ranges with erosional cliffs and scarps, these parallel the strikes of the forereef beds faithfully (Jennings and Sweeting 1963).

The narrow band of the reef proper is solutionally etched as a shallow trench below the forereef and backreef facies. Wherever either of these facies (though more commonly the lagoonal) tends to impurity, rounded relief with a normal valley system occurs

36 Deepdene, Augusta, Western Australia. Gorge due to building of Pleistocene dune ridge athwart river course. Consolidated to aeolian calcarenite.

with only minor solution sculpture, some caves and springs indicative of karst tendencies. Where pure, both facies give rise to distinctive, rugged karst relief (Pl. 25) evolving from 'giant grikeland', intricate karst corridor and fissure cave terrain, through an angular box-valley stage, to a pedimented tower karst. Differences between forereef and backreef karsts are obvious but minor, largely arising from the depositional dip of the former and the horizontal attitude of the latter, and from other lithological details such as the megabreccias of the forereef (Playford and Lowry 1966).

The Chillagoe karst of north Queensland, though subject to basically the same processes in similar climate of modest rainfall and long dry season, differs geomorphically through a different structural context. The rocks here dip steeply, with biohermal lenses of limestone interbedded with impervious rocks. Pedimentation has produced a tower karst only, each lens encroached on all sides to yield a tower or group of towers, riddled with caves and wildly etched by surface solution (Daneš 1916; Jennings 1966).

HOLOKARST AND MEROKARST

Some of these differences due to structure are illustrative of Cvijić's classic distinction between *holokarst* and *merokarst* in Yugoslavia. The holokarst is found in the extensive coastal belt of the Dinaric karst which has great thicknesses of pure limestones reaching deep below sea level and high above it (Gams 1969). Here the full gamut of mid-latitude karst forms and drainage, including erosional poljes, occur; karst processes dominate morphogeny without significant vertical or horizontal boundary effects.

Inland the karst rocks are thinner and interrupted in outcrop by impervious rocks which greatly influence the karst. Karst here is less complete, particularly with regard to poljes, only that of Pester being typical, and normal valleys are more frequent. This inland area was that originally designated as merokarst by Cvijić (1893).

Later, however, Cvijić (1918, 1925) transferred this term to karsts still less elaborate in character where the rocks are lithologically even more unfavourable to the evolution of some karst attributes, especially through the accumulation of residual covers on the karst rocks. Surface solutional sculpture is weak, but not subsoil sculpture; dolines form chiefly in the covers rather than in bedrock; poljes are lacking; caves and potholes may be poorly developed also, whereas integrated valley systems are frequent. These characteristics in the main apply to the Cretaceous Chalk country of England and France, the Jurassic dolomites and limestones of the Franconian Jura, and to other central European karsts according to Cvijić (1960). The inland Dinaric karst he then placed in a transitional group with such karsts as those of the Causses of central France and of the Alps. It may be that there is every gradation from highly developed karst to its most modest manifestation over a range of structural frameworks with no critical thresholds for geomorphic response.

STRUCTURE IN TROPICAL HUMID KARST

Variety of structural relationships is well illustrated by limestone towers in Sarawak (Wilford and Wall 1965; Fig. 62). Their margins may be defined by the top and bottom of vertically dipping limestone interbedded with weaker rocks(a). Where there are sharp changes in facies, the cliff may be not far removed from where the limestone passes laterally into other rocks(b). The facies change may be a less marked one within the karst rocks; some of the limestone flats are developed in bedded, less pure calcarenites and calcirudites whereas the hills are made of very pure, massive, recrystallised coral-algal biolithite(c). Some sides run along faults juxtaposing karst and impervious rocks(d), and igneous dykes cut up other limestone bodies into towers separated by corridors(e) (Pl. 37). Nevertheless many margins are not structurally controlled and lie within the one limestone facies; these are the product of lateral solutional undercutting from alluvially veneered limestone plains(f) or of sapping at basal contact with sloping igneous surfaces(g).

In northern Puerto Rico, five calcareous formations strike west and the karst has a similar latitudinal pattern of different associations of karst forms (Monroe 1968). Cockpit karst is associated with pure, massive limestone, whereas mogote karst requires similar limestone which has had a young surficial cover of sands

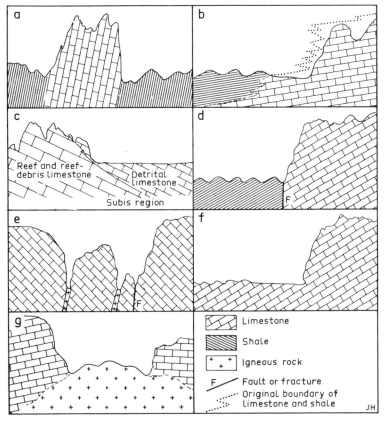

62 *Types of limestone tower in Borneo. After Wilford and Wall 1965.*

and sandy clays, now dropped by solution to veneer the plains between the towers. Less pure, well bedded limestones give rise to zanjones country, shallowly joint-trenched karst. Alternation of mechanically strong and weak limestone gives rise to deep doline karst with which are found the greatest cave developments and the natural bridges. Although the southern part of the karst has been uplifted more than the northern (Gerstenhauer 1964), the climate is uniform and it is lithology which chiefly differentiates karst in this region.

STRUCTURE WITHIN CAVES

Where limestones or dolomites are nearly horizontal, cave plans often show much joint control (Fig. 50), whereas in cross and longitudinal sections, bedding becomes important in delineating roof and floor, with joint planes influencing walls, e.g. in many caves of Craven, England and Clare, Ireland.

When dip becomes substantial, strike directions frequently become important in plan whilst cross-sections tend to be triangular with a sloping bedding plane roof. Along the strike, level passages occur in the bedding but down the dip bedding and jointing alternate in control to yield 'up and down' profiles developed phreatically until destroyed or evaded in vadose development, e.g. in Mendip, England (Ford 1965b; Fig. 63). Deep potholes such as Gouffre Berger have oblique passages down the dip alternating with gravity-controlled vertical pitches.

When beds dip more or less vertically, the cave plan may be dominated by the strike of the beds and the walls are frequently bedding planes, e.g. in Mair's Cave, Flinders Ranges, South Australia (Fig. 64). Joints governing breakdown make ceilings irregular where these are not planed by water, e.g. in the Dip, Wee Jasper, N.S.W.

Recently Bögli (1969) has argued that the role of joints in cave formation has been exaggerated. Because of their greater continuity, bedding planes are more important in the initiation of phreatic caves but the elliptical passages developing in them misleadingly lose that form by secondary breakdown (incasion) along joints (Fig. 53g). Vadose systems, however, will normally originate in open joint conditions which then can prevail in structural guidance. The world's longest caves do not owe much to joints in their passage directions as Bögli himself shows for the Hölloch Cave in Switzerland and Deike (1967) for the great Kentucky caves.

The different attitudes of cavernous rocks are part and parcel of larger tectonic structures, but closer links can exist between the latter and caves through their guidance of solution. Narrengullen Cave near Yass, N.S.W., lies along the axial trough of a syncline at the base of limestone beds surviving only as an inlier in this tectonically low position. Some caves have developed along faults; in Craven, Rift Pot, Long Kin East Pot, Hull Pot and Meregill seem to be of this nature (Myers 1948), though more general claims of fault control in this area are invalid. Faults do not always favour cave development, possibly because of mineralisation or too great a tendency to collapse; Glennie (1950) points out how Ogof Ffynnon Ddu in South Wales has two main systems separated by a fault zone where passages are few.

The details of caves generally reflect structure with a faithfulness as great as that of desert weathering. In northwest Clare, chert bands form false floors and narrow shelves on passage walls, which also result from modest differences in limestone lithology

37 *Joint controlled karst corridor in tower karst in Kinta valley, Perak, West Malaysia*

(Ollier and Tratman 1969). Calcite veins sometimes are resistant to solution and may project as boxwork (Bretz 1942); alternatively voids along the centre of the veins promote solution and

38 Rectangular cross-section and blockpile in horizontal Ordovician limestone in Exit Cave, Tasmania. Photo by R. Curtis.

then they are recessive (Ollier and Tratman 1969). In Craven, bedding-plane passages are frequently associated with thin shale bands in generally pure limestones (Sweeting 1950). In Black

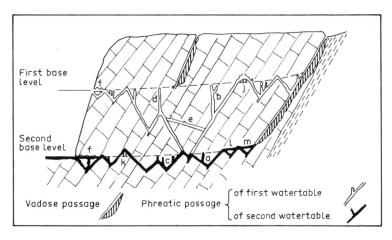

63 Influence of steep dip on Mendip cave development. After Ford 1965b.

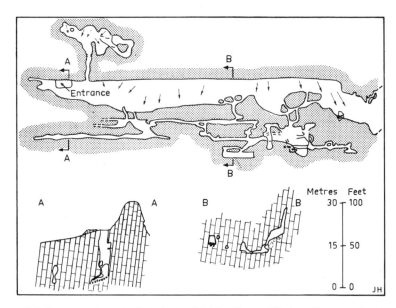

64 Influence of vertical beds in Mair's Cave, Flinders Ranges, South Australia. After Cave Expl. Grp S. Australia survey.

Range Cave, Cooleman Plain, N.S.W., felsite dykes cause doglegs when the stream passage meets them. These few instances must suffice to point to an aspect of cave geomorphology of infinite variety.

HYDROGEOLOGIC SYSTEMS

Not only are the details of individual caves responsive to structure but the whole nature of cave development and the associated underground drainage. This is illustrated by the seven types of hydrogeologic systems that White (1969) has conceived for karst of low to moderate relief (Table 7, Fig 65). Karst rock lithology, depth of karst in relation to river talwegs, relationship to impervious formations and attitude, are the genetic factors incorporated, and all these are essentially structural in the sense adopted here (see overleaf).

TABLE 7 **Hydrogeologic systems in low to moderate karst relief (White 1969)**

I. **DIFFUSE FLOW**
In impure limestones and coarse dolomites. Little surface karst. Random small solution cavities along joints and bedding planes but well connected. Darcy's Law applies. Marked watertable. Large number of small springs. Deep flow.

II. **FREE FLOW**
Thick, massive soluble rocks. Well integrated branchwork caves. Large turbulent flows under gravity at low gradients. A few big springs.

   a. **DEEP**
Karst reaches below river valleys.

      1. **OPEN**
Karst reaches surface. Intakes through many dolines and stream-sinks. Much clastic load. Main waterfilled caves just below river level. Abandoned tubular cave fragments above, sediment-choked.

      2. **CAPPED**
Intake from edge of impervious cap down shafts. Large nearly horizontal caves.

   b. **PERCHED**
Impervious basement above river valleys. Shallow flow paths. Cave streams often free-surface, small water storage.

      1. **OPEN**
Karst reaches surface. Intake through many dolines. Much clastic load. Short cave segments.

      2. **CAPPED**
Vertical shafts round edge of cap and lateral inflow. Long integrated caves.

III. **CONFINED FLOW**
Flow restricted by beds.

   a. **ARTESIAN**
Under inclined cap, hydrostatic pressure forces water deep. Slow flow. 3-D network caves in joints, inclined down beds.

   b. **SANDWICH**
Thin karst rocks between perching and capping impervious rocks. Horizontal 2-D network caves in joints. Small diffuse recharge from above. Backflooding from rivers.

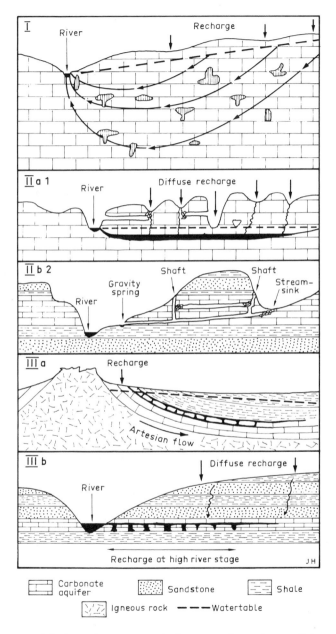

65 *Types of hydrogeologic systems in low to moderate karst relief according to White 1969*

XI

HISTORICAL GEOMORPHOLOGY OF KARST

THE KARST CYCLE

Soon after W. M. Davis's 'cycle of erosion' made impact in Europe, theoretical conceptions of a karst cycle of erosion appeared from those of E. Richter in 1907 onwards.

The most universally based scheme is that of Grund (1914; Fig. 66), which may be summarised as follows. As soon as the land is raised above the watertable, dolines develop at points favouring solution. In youth these are irregularly scattered over the initial surface and most European karst is at this stage. Youth ends when dolines increase in numbers and size so much as to destroy all vestiges of the initial surface and are separated by ridges only. Then the more favoured dolines expand at the expense of neighbours, eliminating intervening ridges. Uvalas develop in this way. Flat floors appear in the depressions and

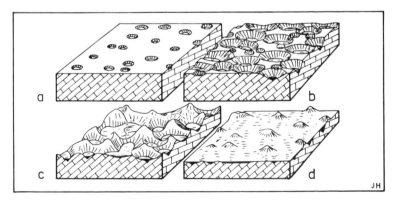

66 Karst cycle according to Grund 1914

cone-shaped hills result from ridge destruction. This is maturity and Grund mistakenly thought that this concept corresponded to the cockpit karst of Jamaica and Java. It is in fact much more like some high mountain karst in New Guinea (Jennings and Bik 1962).

The whole surface is progressively lowered until the depression floors reach the watertable when gentle streams run across them. At this old age stage there is also much cave collapse; gorges and collapse dolines result. Corrosion plains with scattered residual hills are the end point. Poljes are regarded as extraneous tectonic features which can introduce old age forms at younger stages. Though later work has sustained Grund's setting apart of uvalas and poljes, his scheme is vitiated by the amalgamation of temperate and tropical forms into one framework with no regard to climatic factors.

Cvijić (1918) restricted himself to features found in the Dinaric karst in his scheme and he also assumed the simplifying conditions of a thick limestone mass sandwiched between impervious formations but lying entirely above sea level (Fig. 67). As the impervious cover is stripped off, the limestone inherits a normal surface drainage and a valley system develops. Then follows progressive loss of surface streams as solution opens up ways for infiltration and engulfment. Dolines begin to form and tectonic poljes may be present. Eventually all surface drainage has been lost at maximal karst development, dolines occurring on interfluves as well as along the valleys. Elaborate cave systems feed risings around the karst margins and around the poljes, which may, however, be punctured by uvalas now.

Thereafter the karst is destroyed. Normal valleys reappear through steepheads recessing the margins, through cave collapse and through floods planing the polje floors afresh at lower levels, leaving hums here and there. Surface erosion breaches dolines and uvalas, though poljes continue to develop. Finally the impervious basement is widely exposed and only scattered hums remain of karst relief.

Cvijić's scheme can also be criticised. For instance, fluvial drainage may never attack the top surface of the limestone since subsidence and collapse dolines may develop in the overlying rocks before it is exposed. However, it does not claim universality, though Cvijić, like Davis, suffered from his disciples' mis-

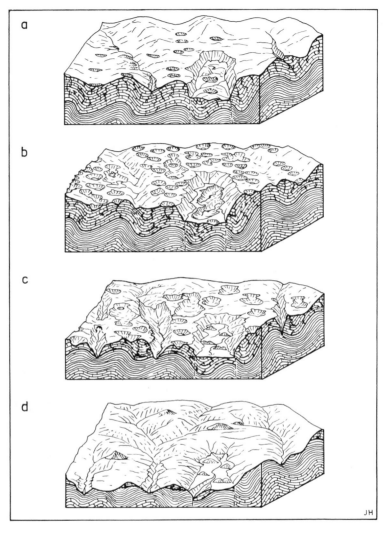

67 *Karst cycle in thick limestone between impervious formations and above sea level according to Cvijić 1918*

representing his ideas (e.g. Sanders 1921). It retains value as a theoretical construct against which to match regional instances, and furnishes the basis for a restatement by Gèze (1965) with the speleological aspects elaborated.

The recognition that karst development may have a number of separate rationales in different climates led investigators to hypothesise evolutionary schemes for particular climato-morphogenic systems, e.g. Lehmann's interpretation of the south-central Javanese karst (p. 187).

However, as Jennings and Sweeting (1963) found in the tropical semiarid karst of the Limestone Ranges of West Kimberley, the problem is complicated by climatic changes. Is the sequence of relief forms evident in these ranges an active product of the present alternation of a short but intensely rainy wet season and a long dry season? Or is it the result of historic events, a tropical humid climate followed by a tropical semiarid or even arid one (Jennings 1969)?

Similar problems arise in better investigated areas such as the Dinaric karst itself. In certain parts, e.g. in the Beljanica Mountains (Gavrilović 1969), hums are set so close as to resemble tropical conekarst. Lateral solutional undercutting has been proposed as the cause of mid-latitude poljes as it has for corrosional plains of tropical humid karst, though minor related features such as tropical cliff-foot caves are lacking in Yugoslavia. It has therefore been suggested that these features are relict from a Pliocene subtropical humid climate (Roglić 1954; Rathjens 1954). Further, Melik (1955) considers that later cold periods in the Pleistocene had the effect of halting or slowing down polje development so that the Tertiary forms survived. In these periods slopes became more active; frost wedging increased, solifluction and landslips moved masses of regolith downslope, and rivers alluviated polje floors more abundantly. Thick layers of gravel, sands, and clays accumulated and some of the underground passages were obstructed. Some highlying poljes are filled to the level of the lowest col and there has been no extension of their floors since this event (Rathjens 1960). According to Roglić (1964b), aggradation of glacifluvial, as well as periglacial fluvial deposits has led to the enlargement, even the creation of polje floors in many Yugoslav instances and in a few the associated blocking of ponors led to Pleistocene lake clay accumulation. Gams (1969), however, maintains that the effects of the cold periods were not solely constructional; increased amounts of runoff from melting snow and ice, accentuated by frozen ground, were directed against marginal limestones and poljes enlarged in this way also.

Cold climates intervene in other ways in karst development. The limestone pavements of Craven, Western Ireland and the Alps belong here (Williams 1966a). These horizontal or sloping platforms of bare karst are due to exposure of well bedded, hard limestone to strong Pleistocene glacial erosion. That erosion has removed weathering mantles and much of the preglacial solutional relief. Most of the minor solution sculpture found on them has developed since the ice retreated. Some variation in lithology seems necessary; more resistant, pure and compact beds form the pavements whereas weaker, impure or poorly cemented beds or shale interbeds are found at the foot of the risers between the platforms. Apart from large, level platforms where both beds and ground are horizontal, pavements occur on moderate slopes where they may be inclined pavements if dip and slope coincide or stepped pavements (*Schichttreppenkarst* of Bögli 1964) if the dip of the beds is with the slope but less than it. Past periglacial phases also leave other effects in the present landscape of temperate karst than their influence on polje development as Tricart (1955) showed in respect of the Causses of France.

Much more extensive survival of well developed tower karst from Tertiary times, in this case lower Tertiary, is claimed for South China by Silar (1965). Gellert (1962) provides stratigraphic evidence for the age in the form of Mio-Pliocene lavas and sediments overlying some of the associated plains. In the higher and more northerly tower karsts of the Yunnan and Kweichow plateaux, inheritance from the warmer Tertiary climate gives an apparent conflict between the nature of the karst and the present climate.

Another contrast leading to inference of inheritance is the presence of inactive cave systems in semiarid and arid karsts as in the Nullarbor Plain in Australia (Jennings 1967c). They are thought to have formed during phases of greater effective, if not absolute precipitation in Pleistocene cold periods. The time of formation of Carlsbad Caverns and other caves in the semiarid Guadalupe Range of New Mexico has been cast farther back still into the wetter Pliocene by Bretz (1949) and Horberg (1949). They argue that the caves are unrelated to present topography and formed by deep phreatic action beneath a summit planation surface, subsequently veneered by Pliocene gravels. Contrarily, Moore (1960) recognises three levels of shallow phreatic development in Carlsbad Caverns which would indicate that at least most of the development took place after uplift and dissection of the summit surface, probably in the Late Pliocene and Early Pleistocene, when, however, conditions were still wetter than at present. There are coarse fluvial sediments present from a late stage in the cave evacuation which are quite incompatible with present climate.

In caves in the Transvaal, alternation of deposition of red sands and formation of speleothems has been related to drier and wetter phases in the Pleistocene (King 1951). The red sands are partly angular grains of local provenance only and partly well sorted and rounded grains of quartz and also of minerals foreign to the vicinity. The latter fraction is regarded as an aeolian component associated with more arid climates than at present which is one of moderate speleothem precipitation. In other Transvaal caves, however, periods of speleothem formation, which alternate with phases of filling with waterborne red earth, are interpreted as the drier parts of arid-humid climatic oscillations (Marker and Brook 1970).

TECTONIC MOVEMENTS AND CHANGES OF BASE LEVELS

Not only do karsts reflect episodes of climatic history, they often register tectonic events and changes of base level more completely than do other terrains: the underground partakes of the effects of these vicissitudes as well as the surface. This is illustrated by the 300 m fault scarp on the western side of Mt Hoyo in the eastern Congo, which is penetrated by caves at various levels (Ollier and Harrop 1963). Although opening from the face of this high cliff, the caves are phreatic in nature, though they have subsequently been modified vadosely, especially near their entrances. The phreatic phase is incongruous with their present position and they must antedate early Pleistocene faulting. Cave survival through this tectonic phase is remarkable.

General uplift leads to rejuvenation which can be recorded both below as well as above ground as Sweeting (1950) has shown for Craven. Here the Carboniferous Limestone and part of the overlying, mainly impervious Yoredale Beds are truncated by an

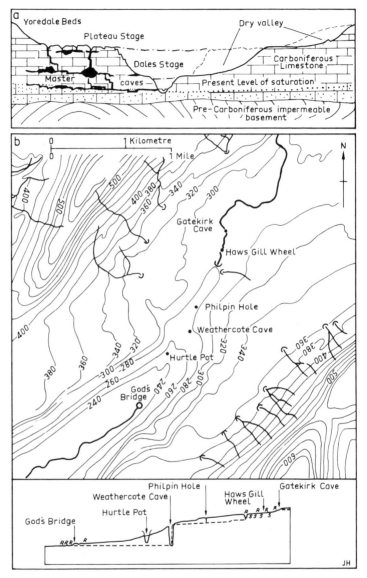

68 (a) *Scheme of karst evolution in Craven, England. After Sweeting.*
   (b) *Cave development at rejuvenation head on River Greta, Craven,
   between Plateau and Dales Stages of valley development. After
   Warwick 1960.*

erosion surface at about 370 m (Fig. 68). Close beneath this surface horizontal caves formed and valleys graded to it. In a first rejuvenation the valleys in limestone became dry and underground drainage descended to greater depths, deep potholes being formed and vertical pitches added to earlier cave systems. Then in a stable base level stage, the Dales Stage, straths were fashioned along many valleys 70-100 m lower down and so also horizontal cave passages at about 280 m, feeding surface streams at the strath level. A second rejuvenation ensued with further incision of the major rivers to grade to the Craven Lowlands, leaving more tributary valleys hanging and dry. Further vertical development took place in the caves and passages at the 280 m level were enlarged into large chambers by collapse after emptying of water. A final stillstand has caused horizontal 'master' caves, largely waterfilled, at present rest levels at about 200 m.

Not all caves in the area are old and complex in their history. Residual hillmasses of the Yoredale Beds continue to recede through surface erosion, and, as their margins retreat, fresh potholes develop at streamsinks on the still extending 370 m surface; some streamsinks are of Holocene age.

Warwick (1960, 1962) has stressed the importance of the effects of successive phases of rejuvenation on British karsts generally. In particular, rapid lowering of water levels underground near to rejuvenation heads in limestone valleys leads to sinking of rivers and cave formation (Fig. 68). This is particularly liable to happen where nickpoints pass up incised meanders, short cuts developing through meander cores in these circumstances (cf. p. 102).

Denudation chronology in karst is likely to have a more precise record because of the perfection of corrosion plains (Pl. 19) compared with those formed in other ways and because of the degree of immunity from valley erosion which they possess through partial deflection of water underground.

The relict karst margin plains of the Dinaric karst have long attracted attention (Krebs 1929; Cholley and Chabot 1930; Kayser 1934) because of their high degree of planation evident after dissection and the sharp breaks they make with the high plateaux. These plains occur both along the inland margins of the karst and along the Adriatic coast, e.g. the Karlovac-Sluin, Kistanje, and Cetina plains. Rivers, such as the Krka and Čikola in the case of the Kistanje plain, meander in trenches through the surfaces. Hums rise from their surfaces and dolines occur especially near to the river gorges. Morawetz (1967) compares these plains at elevations of 200-300 m with the corrosion plain of the lower Neretva valley (p. 143). He argues that higher surfaces were fashioned in a similar way prior to a negative movement of base level which raised them high above water rest levels and began their protracted slow destruction.

Accordance of summit levels in tropical conekarsts has been interpreted as evidence of planation surfaces from which the fields of residuals have been carved. For example near Baisha southeast of Kweilin in southern China, limestone towers betray a storeyed relief with three surfaces represented by accordance at different levels. Gellert (1962) considered there had been three corrosion plain stages separated by uplift and rejuvenation (Fig. 69). High up in individual towers there are cliff-foot caves and through-caves also witnessing this storeying.

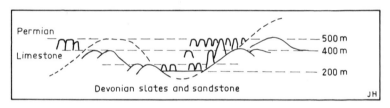

69 Storeying in tower karst, Baisha, Kweilin, China. After Gellert 1962.

Just as the interpretation of summit accordances and upland plains in impermeable rocks as remnants of former erosion surfaces partially destroyed by rejuvenation has been subject to latter-day criticism in favour of other explanations, for example in favour of dynamic equilibrium, so also have special objections of similar purpose arisen within the realm of karst. Thus for the Jura, Aubert (1969) maintains that doline deepening is eventually stultified by the accumulation of insoluble fines within them and erosion is then directed laterally. In this way relief is flattened, especially because the more open jointing along anticlinal axes favours the reduction of tectonic relief. It is his view that the high plateaux of the Jura, previously interpreted as relics of a Tertiary peneplain, are in fact relief conformal to tectonics but attenuated, if not completely planed, by karst surface corrosion which has operated and still is operating laterally rather than vertically.

Whether some of the rejuvenations discussed above are the product of tectonic uplift or of large eustatic shifts of sea level itself enters a debate not central to the theme of this book (see Bird 1968 and Twidale 1971). Nevertheless there is no doubt that glacioeustatic sea level changes have affected karsts, most obviously through the alternating exposure and drowning of the continental shelves. Where there are surface streams in coastal karst, they deepen their valleys during glacial low sea levels and these are drowned in the interglacials, finally by the Flandrian transgression in the Holocene. In the Dinaric karst this is evident in the Gulf of Sibenik which consists of the drowned lower valleys of the Krka and Čikola. The Kistanje erosion surface extends over some of the Adriatic islands and some of the sea channels between them, the *kanali,* are probably drowned Pleistocene valleys. However, little has been done to see whether such drowned valleys possess special karst characteristics. Certain of the 'calanques' of the limestone coast of Provence are coastal inlets of a gorge-like nature and Chardonnet (1948) explains them as due to the formation by freshwater solution of caves reaching below sea level followed by roof collapse and eventual destruction of the seaward barrier to allow drowning without any change in sea level. However, no case for rejecting the alternative explanation of Flandrian flooding of collapsed caves formed during low sea level times is given.

Some submarine springs are a result of deepening of underground circulation during glacial low sea levels and subsequent submergence (p. 79). Other submarine features of which the same may be said are the large closed depressions on the Florida shelf (G. F. Jordan 1954). The ones surveyed vary between 40 and 160 m in depth and between 900 and 2200 m across; their tops are at −300 to −500 m and lie on a fault block above a fault scarp. Despite these depths and the unstable tectonic context, it is likely that eustatic low stands of sea level have contributed to their formation. R. H. Jordan (1950) previously argued that vadose action during Pleistocene low sea levels had worked in conjunction with deep phreatic circulation to fashion the deep, water-filled dolines of the Florida peninsula itself.

Well known because of its spectacular landscape, the archi-

39 Drowned tower karst in Langkawi Islands, Kedah, West Malaysia

pelago of Vung Ha Long Bay in north Vietnam can be described as a tower karst sitting in a shallow sea. The same is true of parts of the Langkawi Islands of Malaya (Pl. 39). Some of the Vung Ha Long islands enclose lakes of sea water reached by boats through caves. Though active marine notches undercut these towers now, it is doubtful whether the relief as a whole has been produced at present sea level. Marine processes can produce only a few stacks in front of a shoreline retreating from marine attack but cannot fashion such extensive archipelagoes on their own. Some of the Vung Ha Long towers have bases at −20 m. Therefore change of base level is involved. Silar (1965) postulates tectonic downwarping (followed by uplift because of emerged marine terraces on the nearby mainland shore and of caves in the towers at +14-19 m). However, adequate positive and negative movements due to glacioeustatism are known to have occurred which could explain all the phenomena without tectonic deformation.

Reference has been made to the effects of glacioeustatic shifts on cave history in Bermuda, ending in the drowning of some caves and their speleothems. Caves above sea level sometimes reveal effects of alternating sea level. In Auckland, New Zealand,

Kairimu Cave lies along a shallow syncline in Lower Oligocene limestone which was uplifted in Late Miocene times (Barrett 1963). Its oldest passages were then initiated phreatically and subsequently much enlarged by vadose action. High sea level in Late Pliocene or Early Pleistocene time flooded the cave which opens at +100 m, and deposited silt and clay with a dating microflora. Some small passages have developed since. As this part of New Zealand is relatively stable tectonically, the stage of marine drowning of this cave could be of eustatic origin only.

A miniature relevant example is provided by Narrengullen Cave well inland in southern New South Wales which has been affected by water level oscillations in Burrinjuck Reservoir, an artificial storage which laps to its exit in high stages. This has caused alluviation of the lower end of the cave. Big flows and sediment loads tend to correspond with high lake levels; reduced river flows during low lake stages cannot readily remove the intervening deposition.

COMPLEX EVOLUTION

The artificiality of separate discussion in this and previous chapters of the operations of individual factors differentiating karst has been made patent by inevitable cross-references. Most karst evolution is complex, both on the surface and underground.

In Clare, Ireland, Visean limestone underlies Namurian sandstone and shales, which mainly form gentle but higher land draining on to the karst. The shales were breached in the late Tertiary when large depressions were initiated (Williams 1970). Though they developed further in warm Pleistocene intervals, their catchments were gradually removed by glacial erosional stripping of the shales and they ceased to develop, though one, the Carran depression, had become a polje (Sweeting 1953). Bevelled spurs, hill-top summits and river nickpoints aggregate into a series of levels, which Sweeting (1955) regarded as late Tertiary erosion surfaces, but Williams (1970) maintains are largely due to structural guidance of Pleistocene glacial erosion. Some closed depressions are simply hollows in the surface of glacial till on top of the limestone. Glacial erosion produced limestone pavements, destroying previous minor solution sculpture so that many grikes and other small forms now present are largely postglacial.

Despite this polygenetic history, Ollier and Tratman (1969) find the evolution of the caves of a substantial part of the area to be a simple one confined to the Postglacial. These young caves are active branchwork caves at shallow depth, dominantly due to vadose solution with sharp fluctuations of cave river levels. Only a very few are conceded to have a longer history. The caves follow the lines of dry valleys, which have been cut in the Post-glacial with removal of moraine along them. The ice probably retreated about 15,000 B.P. from the area so that the caves have a short and simple history in contradistinction to the surface landforms.

This is a valuable illustration of the important principle that surface and underground forms can get quite out of step in their development. The converse relationship to the one just described may be even more common, namely that cave systems can develop over long periods of time without having much effect on surface topography as Aubert (1969) maintains with respect to much of the Jura Mountains, where, for instance, collapse dolines are very rare.

The Mendip Hills in southwest England are complex in their whole karst evolution (Ford and Stanton 1968). Valleys cut in the Carboniferous Limestone were buried by Triassic conglomerate and deposits of the same age filled some caves. Subsequently the valleys have been partially exhumed so that there are Triassic elements in the landscape. The Mendip plateau surface is due to Pliocene erosion with drainage oriented by Miocene doming, though parts of a Jurassic unconformity survive through coincid-ence with the Pliocene level. Valleys were incised into this surface and erosional benches are recognised in them. These events are also registered in the development of the caves which accompanied stream sinking. The valleys are now dry except on impervious inlier monadnocks and below the major risings. Cutting of the valleys is largely attributed to permafrost causing resumptions of surface drainage during the Pleistocene cold periods. Cheddar Gorge is due entirely to surface incision and Ebbor Gorge is due to the destruction of a cave by surface incision during a cold period. In the cold periods the caves were chiefly subject to calcite deposition and cave enlargements took place during the intervening interglacials when there was no ground ice to minimise under-ground circulation. However, these phases did not lead to stoping to the surface and dolines along the dry valleys are the expression of surface solution where gradients are gentle. Large shallow basins at the heads of former surface drainage were ponds during the periglacial periods, when they collected fills of clay, partly wind-borne. Thus from lower Pleistocene times onward some of the caves have been affected by alternations of rejuvenation and still-stand, and by the impact of successive phases of cold and of temperate climate in inextricable genetic association with the con-temporaneously evolving surface of the karst.

In the light of such contrasts in evolution between two com-paratively well studied karsts, both Carboniferous Limestone plateaux with Hercynian tectonic histories and subject to similar climate today, present knowledge may be inadequate for defining systematic karst types on the total basis of structure, climato-geomorphic processes and historical evolution, although Birot (1954) has attempted this for the Alpine and Hercynian belts of Europe.

XII

PRESENT STATE OF KARST
GEOMORPHOLOGY AND ITS VALUE

Present knowledge of karst is microcosmic of the whole discipline of geomorphology. Because it includes some of the most difficult terrain for movement there is, even qualitative description is still incomplete for many areas of karst, particularly within the tropics (Pl. 35). But these frontiers are being pushed back now with modern aids, though the geomorphologist may be a 'poor relation' riding oil company helicopters. Techniques and equipment for the exploration of practically any kind and size of cave have been elaborated, and diving methods have been modified to permit a degree of direct examination of waterfilled passages; yet many karsts are speleologically almost uninvestigated. Photogrammetry presents special difficulties in karst, requiring more ground checking than in many other landscapes. Sharp variation in rainforest height over conekarst points to the need for sensors other than black-and-white and ordinary colour film for mapping here.

Complexity in much karst relief, with its apparently chaotic nature, has perhaps daunted application of modern morphometric methods in this field. However, recent endeavours along these lines are bringing karst geomorphology more into step with other aspects of the discipline with promising results.

Karst is normally included in texts devoted to structural geomorphology since it is fundamentally distinguished from other topography by the effects of abnormal rock solubility. Nevertheless there has been a tendency to take the nature of the rocks for granted. Recent signs of pursuit of finer lithologic variations in their karst effects keep this branch abreast of similar indications elsewhere in geomorphology.

The equal need to approach karst in terms of the role of different

224

climates in conditioning its development in distinct morphogenic systems was exemplified early in the impetus towards climatic geomorphology between the two world wars. At least the contrasting morphologic regimes of periglacial, hot arid, and hot humid zones with their attendant suites of karst landforms stand forth from a proliferation of proposed climato-morphogenic realms in karst. The readiness with which karst students turned to the climatic approach went with an enduring interest in karst processes; there never was on this side of the discipline quite as much neglect of process as has been associated with the Davisian school of geomorphology. In this connection it would be unjust not to recollect the stimulating attention that Davis himself gave to the analysis of different hydrodynamic circumstances in underground geomorphology. In recent decades the chemical aspects of karst processes, including rates of chemical denudation, have attracted a great deal of effort and practitioners of karst geomorphology can feel reasonably satisfied with their contribution to the surge of activity in quantitative geomorphic process study. That this has led in the last few years to increasing involvement of physicists and chemists in what are essentially geomorphic problems is to be welcomed (Ashton 1966; Roques 1969), even though some matters thereby become insulated from criticism by more generally trained geomorphologists. Glaciology has had this experience.

All these approaches must be brought together in order to reconstruct the historical evolution of individual karsts. In this task there are both special advantages such as the perfection of corrosion plains and special obstacles such as the difficulty frequently met in dating stages of cave development. Many more results must come forth from the different approaches—morphometry, process measurement, correlative sediment study—before it can be resolved whether, in this final stage of synthesis, karst study must remain a nomothetic rather than an ideographic science. 'Hydrogeologic uniqueness' is not infrequently invoked in respect of individual limestone areas (Stringfield and LeGrand 1969a). At the moment the essential point is that karst investigators are participating in all the growing points of geomorphology, if to differing degrees.

This book has been concerned with the purely scientific objectives of karst investigation; it is proper to conclude with a brief indication of its practical advantages.

Questions of drinking water supply have long prompted much activity. Martel's name stands out in this connection, especially with regard to danger of pollution arising from the large conduits and rapid transit common in karst groundwater. In arid and semi-arid zones, karst rocks assume very great importance as aquifers and this has led to much work by FAO on karst hydrology in the environs of the Mediterranean. Not only surface karst is significant in this respect; buried karsts are important aquifers in the United States (Stringfield and LeGrand 1969a). Many karst springs have mineral contents of alleged curative value and give rise to spas.

In times past, karst springs have been used to generate mechanical power, but more recently knowledge of underground drainage has helped in the assessment and manipulation of hydroelectric potential as in the Gouffre Pierre St Martin in the Pyrenees. Dubrovnik in the Dinaric karst has an electricity supply from springs within the city limits. In the neighbourhood of Antalya in southern Turkey the karst drainage system has required deflection for irrigation needs. Understanding of the functioning of poljes in Yugoslavia has permitted engineers to diminish flooding of their cultivable floors, and in Florida and Georgia wells are put down into limestone to get rid of surface drainage. The handling of water resources remains the prime economic spark for karst research.

Mining of phosphate and nitrates from caves is more of historic than present commercial interest. However, karst cavities of past geological periods have been lodging points for other minerals such as alluvial tin, sulphide and other ores of copper, and they have also acted as oil reservoirs. Search for and exploitation of these resources can be pursued more successfully in the light of improved knowledge of karst in all its manifestations.

Engineering in karst country takes risks if its peculiarities are disregarded. The May River dam near Konya in central Turkey is a costly construction unlikely ever to perform its purpose because its catchment, floored with limestone, behaves like a sieve (Pl. 40). There are also problems of dam stability to be reckoned with on limestone foundations. Lowering of water levels in dolomites by pumping from gold mines in the Far West Rand in Transvaal, South Africa, has caused accelerated subsidence of deep soils and regolith into cavities (Brink and Partridge 1965).

40 Subsidence doline in alluvium over limestone in May River dam near Konya, Turkey. Formed since dam was constructed.

This had led to considerable loss with the extreme case of the disaster at West Dreifontein in 1962 when twenty-nine workers lost their lives in sudden subsidence which engulfed a crusher plant. Nowadays in parts of the United States, geophysical sounding is employed over proposed highway routes through cavernous terrain so that thin roofs can be avoided. Better understanding can lead to better management of land.

Caves have long been used as refuges, particularly by outcasts and escapists from society. In Europe during World War II, people fled to caves as a protection from air raids, and in New Guinea evidences of recent cave occupation are ascribed by the New Guineans to the 'taim bilong big pait'. Since that war, consideration has been given in some countries to the potential of caves as refuges for large numbers in times of nuclear warfare. They can also provide cheap storage—in peace as with Roquefort cheese in the Causses of France and in war as with Nazi petrol and oil reserves in Postojna Cave in Slovenia until partisans coming by unguarded distant entrances set fire to the military dumps. Whether cave air has any real therapeutic value for illnesses such as asthma may be uncertain, yet in Hungary and Turkey there are caves employed in this way.

The tourist attraction of well decorated or otherwise impressive caves is a resource in process of wider realisation than ever before as societies become more affluent and better educated. The same is true of spectacular surface features such as natural bridges, tufa dams and gorges which often accompany them. There are great conservation problems arising, with tourism the largest single and fastest growing industry of the world. For example, the placing of hotels requires the greatest of care lest it involve partial ruin of their *raison d'être*. This has happened at various places such as on the magnificent wall of rimstone dams at Pamukkale in western Turkey. When a forest is destroyed, there is still the chance of natural regrowth under protection in a century or two; when the curiosities of the mineral kingdom which decorate caves in such fantastic and beautiful fashion are broken, on the human timescale they are lost forever. In this matter, however, tasteless commercialisation is a less widespread danger than the enthusiastic but often careless behaviour of cavers in unprotected caves.

Financial support from governments and private enterprises for karst investigation and cave exploration is naturally most forthcoming in countries with extensive karst areas such as France. However, karst has such a fascination that even where this kind of support is not available, geomorphologists, whether amateur or professional, will surely find opportunity to contribute their part to the understanding of it. Its economic and social values will be not less for being a by-product of scientific curiosity and sporting endeavour.

BIBLIOGRAPHY

Adams, C. S. and Swinnerton, A. C. 1937. The solubility of calcium carbonate. *Trans. Am. geophys. Un.*, **11**:504-8.

Ashton, K. 1966. The analysis of flow data from karst drainage systems. *Trans. Cave Res. Grp Gt Br.*, **7**:161-203.

Aub, C. F. (in press a). Some observations on the karst morphology of Jamaica. *Congr. 5ᵉ int. Spéléol.*

——— (in press b). The nature of cockpits and other depressions in the karst of Jamaica. *Congr. 5ᵉ int. Spéléol.*

Aubert, D. 1966. Structure, activité et évolution d'une doline. *Bull. Soc. Neuchateloise Sci. Nat.*, **89**:113-20.

——— 1969. Phénomènes et formes du karst jurassien. *Eclogae geol. Helv.*, **62**:325-99.

Balázs, D. 1968. Karst regions in Indonesia. *Karszt-és Barlangkutatás*, **1963-67**:3-61.

Balch, E. S. 1900. *Glacières or Freezing Caverns.* Philadelphia.

Balchin, W. G. V. and Lewis, W. V. 1938. The chalk water table south-east of Cambridge. Pp. 20-4 in *The Cambridge Region*, ed. H. C. Darby. Cambridge.

Barrett, P. J. 1963. The development of Kairimu Cave, Marakopa District, South-West Auckland. *N.Z. Jl Geol. Geophys.*, **6**:288-98.

Bauer, F. 1964. Kalkabtragungsmessungen in den Österreichischen Kalkhochalpen. *Erdkunde*, **18**:95-102.

Bird, E. C. F. 1968. *Coasts.* Canberra.

Birot, P. 1949. *Essai sur quelques problèmes de morphologie générale.* Lisbon.

——— 1954. Problèmes de morphologie karstique. *Annls Géogr.*, **63**:161-92.

Blanc, A. 1958. Répertoire bibliographique critique des études de relief karstique en Yougoslavie depuis Jovan Cvijić. *Mém. Docums. Cent. Docum cartogr. géogr.*, **7**:135-228.

Bögli, A. 1956. Grundformen von Karsthöhlenquerschnitten. *Stalactite*, **6**:56-62.

——— 1960. Kalklösung and Karrenbildung. *Z. Geomorph.-Suppl.*, **2**:4-21.

——— 1961a. Karrentische, ein Beitrag zur Karstmorphologie. *Z. Geomorph.*, **5**:185-93.

——— 1961b. Der Höhlenlehm. *Mem. Rass. speleol. ital.*, **5**(2):11-29.

——— 1964a. Mischungskorrosion—ein Beitrag zur Verkarstungsproblem. *Erdkunde*, **18**:83-92.

——— 1964b. Le Schichttreppenkarst. *Revue Belge Géogr.*, **88**:63-82.

——— 1969. Neue Anschauungen über die Rolle von Schichtfugen und Klüften in der karsthydrographischen Entwicklung. *Geol. Rdsch.*, **58**:395-408.

Bretz, J. H. 1942. Vadose and phreatic features of limestone caverns. *J. Geol.*, **50**:675-811.

——— 1949. Carlsbad Caverns and other caves of the Guadalupe Block, New Mexico. *J. Geol.*, **57**:447-63.

——— 1960. Bermuda: a partially drowned, late mature Pleistocene karst. *Bull. geol. Soc. Am.*, **71**:1729-54.

Brink, A. B. A. and Partridge, T. C. 1965. Transvaal Karst: some considerations of development and morphology, with special reference to sinkholes and subsidence on the Far West Rand. *S. Afr. geogr. J.*, **47**:11-34.

Brown, E. H. 1969. Jointing, aspect and orientation of scarp-face dry valleys, near Ivinghoe, Buckinghamshire. *Trans.Inst.Brit.Geogr.*, **48**:61-74.

Burdon, D. J. and Safadi, C. 1963. Ras-el-ain: the great karst spring of Mesopotamia. *Jnl Hydrol.*, **1**:58-95.

Burke, A. R. and Bird, P. F. 1966. A new mechanism for the formation of vertical shafts in Carboniferous Limestone. *Nature, Lond.*, **210**: 831-2.

Burns, K. L. 1964. *Geological Survey Explanatory Report Devonport.* Tasmania Dept. Mines, Hobart.

Butcher, A. L. and Railton, C. L. 1966. Cave surveying. *Trans.CaveRes. Grp Gt Br.*, **8(2)**:1-37.

Caumartin, V. and Renault, P. 1958. La corrosion biochimique dans un réseau karstique et la genèse du mondmilch. *Notes biospéol.*, **13**:87-109.

Chandler, R. H. 1909. On some dry chalk valley features. *Geol. Mag.*, **6**:538-9.

Chardonnet, J. 1948. Les calanques provençales, origine et divers types. *Annls Géogr.*, **57**:289-97.

Chevalier, P. 1944. Distinction morphologique entre deux types d'érosion souterraine. *Revue Géogr. alp.*, **32**:475-92.

Cholley, A. and Chabot, G. 1930. Notes de morphologie karstique. *Annls Géogr.*, **39**:270-85.

Cleland, H. F., 1910. North American natural bridges, with a discussion of their origin. *Bull. geol. Soc. Am.*, **21**:313-38.

Coleman, A. M. and Balchin, W. G. V. 1959. The origin and development of surface depressions in the Mendip Hills. *Proc. Geol. Ass.*, **70**: 291-309.

Coleman, J. C. 1945. An indicator of waterflow in caves. *Geol. Mag.*, **82**:138-9.

Collingridge, B. R., 1969. Geomorphology of the area. Pp. 42-58 in *The Caves of Northwest Clare, Ireland*, ed. E. K. Tratman. Newton Abbott.

Conrad, G., Gèze, B., and Paloc, H. 1968. Phénomènes karstiques et pseudokarstiques du Sahara. *Congr. 4ᵉ int. Spéléol.*, **3**:411-16.

Corbel, J. 1952. A comparison between the karst of the Mediterranean region and of northwestern Europe. *Trans. Cave Res. Grp Gt Br.*, **2**:1-26.

——— 1954. Les phénomènes karstiques en climat froid. *Erdkunde*, **8**: 119-20.

——— 1958. *Les karsts du Nord-Ouest de l'Europe.* Inst. Et. Rhod. Mem. et Doc., **12**.

——— 1959a. Érosion en terrain calcaire. *Annls Géogr.*, **68**:97-120.

Corbel, J. 1959b. Les karsts du Yucatan et de la Floride. *Bull.Ass.Géogr. France*, **282-3**:2-14.

―――― 1960. Nouvelles recherches sur les karsts arctiques Scandinaves. *Z.Geomorph.-Suppl.*, **2**:74-80.

Cramer, H. 1933. Höhlenbildung im Karsten. *Petermanns geogr. Mitt.*, **79**:78-80.

―――― 1941. Die Systematik der Karstdolinen. *Neues Jb. Miner. Geol. Paläont.*, **85**:293-382.

Cullingford, C. H. D. 1962. *British Caving*. 2nd ed. London.

Curl, R. L. 1958. A statistical theory of cave entrance evolution. *Bull. natn. speleol. Soc.*, **20**:9-22.

―――― 1960. Stochastic models of cave development. *Bull. natn. speleol. Soc.*, **22**:66-76.

―――― 1966a. Scallops and flutes. *Trans. Cave Res. Grp Gt Br.*, **7**:121-60.

―――― 1966b. Caves as a measure of karst. *J. Geol.*, **74**:798-830.

Cvijić, J. 1893. Das Karstphänomen. *Geogr. Abh.*, **5**:217-329.

―――― 1918. L'hydrographie souterraine et l'évolution morphologique du karst. *Revue Géogr. alp.*, **6**:375-426.

―――― 1925. Types morphologiques des terrains calcaires. Le Holokarst. Le Mérokarst. Types karstiques de transition. *C.r.hebd. Séanc. Acad. Sci.*, *Paris*, **180**:590-4, 757-8, 1038-40.

―――― 1960. *La géographie des terrains calcaires*. Académie serbe des Sciences et des Arts monographies, **141**.

Daneš, J. V. 1908. Geomorphologische Studien im Karstgebiete Jamaikas. *9ᵉ Int. géogr. Congr.*, **2**:178-82.

―――― 1910. Die Karstphänomene im Goenoeng Sewoe auf Java. *Tijdschr. K. ned. aardrijksk. Genoot*, **27**:247-60.

―――― 1916. Karststudien in Australien. *Sber. K. böhm. Ges. Wiss.*, **7**:1-75.

Davies, J. L. 1969. *Landforms of Cold Climates*. Canberra.

Davies, W. E. 1949. Features of cave breakdown. *Bull. natn. speleol. Soc.*, **11**:34-5.

―――― 1960. Origin of caves in folded limestone. *Bull. natn. speleol. Soc.*, **22**:5-18.

Davis, W. M. 1930. Origin of limestone caverns. *Bull. geol. Soc. Am.*, **41**:475-628.

Deike, G. H. 1960. Origin and geologic relations of Breathing Cave, Virginia. *Bull. natn. speleol. Soc.*, **22**:30-42.

―――― 1967. The Development of Caverns in the Mammoth Cave Region. Ph.D. thesis, Pennsylvania State University.

Douglas, I. 1964. Intensity and periodicity in denudation processes with special reference to the removal of material in solution by rivers. *Z. Geomorph.*, **8**:453-73.

―――― 1965. Calcium and magnesium in karst waters. *Helictite*, **3**:23-36.

―――― 1968. Some hydrologic factors in the denudation of limestone terrains. *Z. Geomorph.*, **12**:241-55.

―――― 1969. *Field Methods of Water Hardness Determination*. Br. Geomorph. Res. Grp Tech. Bull. **1**.

Drew, D. P. and Smith, D. I. 1969. *Techniques for the Tracing of Subterranean Drainage*. Br. Geomorph. Res. Grp Tech. Bull. **2**.

Dunham, R. J. 1962. Classification of carbonate rocks according to depositional texture. *Am. Ass. Petrol. Geol.*, **1**:108-21.

Duplessy, J. C., Labeyrie, J., Lalou, C., and Nguyen, H. V. 1970. Continental climatic variations between 130,000 and 90,000 years B.P. *Nature, Lond.*, **226**:631-3.

Ek, C. 1961. Conduits souterrains en relation avec les terrasses fluviales. *Annls Soc. géol. Belg.*, **84**:313-40.

——— 1964. Notes sur les eaux de fonte des glaciers de la Haute Maurienne. *Revue Belge Géogr.*, **88**:127-56.

——— 1969. L'effet de la loi de Henry sur la dissolution du CO_2 dans les eaux naturelles. Pp. 53-5 in *Problems of the Karst Denudation*, ed. O. Štelcl. Brno.

——— Delecour, F., and Weissen, F. 1968. Teneur en CO_2 de l'air de quelques grottes belges. Technique employée et premiers résultats. *Ann. Spéléol.*, **23**:243-57.

Emig, W. H. 1917. *Travertine Deposits of Oklahoma*. Oklahoma Geol. Surv. Bull. **29**.

Fagg, C. C. 1923. The recession of the chalk escarpment. *Proc. Trans. Croydon nat. Hist. scient. Soc.*, **9**:93-112.

——— 1954. The coombes and embayments of the chalk escarpment. *Proc. Trans. Croydon nat. Hist. scient. Soc.*, **12**:117-31.

Feininger, T. 1969. Pseudokarst on quartz-diorite, Colombia. *Z. Geomorph.*, **13**:287-96.

Fénelon, P. (ed.) 1968. Vocabulaire français des phénomènes karstiques. *Mem. Docums. Cent. Docum. cartogr. géogr.*, **4**:193-282.

Flathe, H. and Pfeiffer, D. 1965. Grundzüge der Morphologie, Geologie und Hydrogeologie im Karstgebiet Guning Sewu, Java (Indonesien). *Geol. Jb.*, **83**:533-62.

Folk, R. L. 1959. Practical petrographic classification of limestones. *Bull. Am. Ass. Petrol. Geol.*, **43**:1-38.

Ford, D. C. 1964. Origin of closed depressions in the central Mendip Hills, Somerset, England. Paper at Karst Symposium, 20th International Geographical Congress, England, 1964.

——— 1965a. Stream potholes as indicators of erosion phases in limestone caves. *Bull. natn. sp leol. Soc.*, **27**:27-32.

——— 1965b. The origin of limestone caverns: a model from the central Mendip Hills, England. *Bull. natn. speleol. Soc.*, **27**:109-32.

——— 1966. Calcium carbonate solution in some central Mendip caves, Somerset. *Proc. speleol. Soc. Univ. Bristol*, **11**:46-53.

——— and Stanton, W. L. 1968. The geomorphology of the south-central Mendip Hills. *Proc.Geol.Ass.*, **79**:401-28.

Ford, T. D. and King, R. J. 1966. The Golconda Caverns, Brassington, Derbyshire. *Trans.Cave Res.Grp Gt Br.*, **7**:81-114.

——— and ——— 1969. The origin of the silica sand pockets in the Derbyshire limestone. *Mercian Geologist*, **3**:51-69.

Frank, R. M. 1971. Cave sediments as palaeoenvironmental indicators, and the sedimentary sequence in Koonalda Cave. Pp. 94-104 in *Aboriginal Man and his Environment in Australia*, ed. D. J. Mulvaney and J. Golson. Canberra.

Frear, G. and Johnston, J. 1929. Solution of calcium carbonate in aqueous solutions at 25°C. *J. Am. Chem. Soc.*, **51**:2082-93.

Gams, I. 1962. Slepe doline Slovenije v Primerjalni Metodi. *Geogr.Zb.*, **7**.

——— 1963. Die Einfluss der Schichtenlage auf die Richtung der Höhlengänge und auf die Querschnitte in die längsten Höhlen Sloweniens. *Congr. 3ᵉ int. Spéléol.*, **2**:215-20.

——— 1965. Types of accelerated corrosion. Pp. 133-9 in *Problems of the Speleological Research*, ed. O. Štelcl. Prague.

——— 1966. Faktorji in dinamika Korozije na karbonatnih Kameninah Slovenskeya dinarskegain alpskegu Krasa. *Geogr. Vestnik*, **38**:11-68.

Gams, I. 1968. Über die Faktoren, die die Intensität der Sintersedimentation bestimmen. *Congr. 4ᵉ int. Spéléol.*, **3**:107-15.
———— 1969. Some morphological characteristics of the Dinaric karst. *Geogrl J.*, **135**:563-72.
Gavrilović, D. 1969. Kegelkarst-Elemente im Relief des Gebirges Beljanica, Jugoslavien. Pp. 159-66 in *Problems of the Karst Denudation*, ed. O. Štelcl. Brno.
Gellert, J. 1962. Der Tropenkarst in Südchina im Rahmen der Gebirgsformung des Landes. *Verh. dt. Geogr. Tags.*, **33**:376-84.
Gerstenhauer, A. 1960. Der tropische Kegelkarst in Tabasco (Mexico). *Z. Geomorph.-Suppl.*, **2**:22-48.
———— 1964. Blatt 3 (Nord-Puerto Rico) des Internationalen Karstatlas. *Erdkunde*, **18**:148.
Gèze, B. 1953. La genèse des gouffres. *Congr. 1ᵉʳ int. Spéléol.*, **2**:11-23.
———— 1958. Sur quelques caractères fondamentaux des circulations karstiques. *Ann. Spéléol.*, **13**:5-22.
———— 1965. *La spéléologie scientifique*. Paris.
Gilewska, S. 1964. Fossil karst in Poland. *Erdkunde*, **18**:124-35.
Glennie, E. A. 1948. Some points relating to Ogof Ffynnon Ddu. *Trans. Cave Res. Grp Gt Br.*, **1(1)**:13-25.
———— 1950. Further notes on Ogof Ffynnon Ddu. *Trans. Cave Res. Grp Gt Br.*, **1(3)**:1-47.
———— 1954a. Artesian flow and cave formation. *Trans. Cave Res. Grp Gt Br.*, **3**:55-71.
———— 1954b. The origin and development of cave systems in limestone. *Trans. Cave Res. Grp Gt Br.*, **3**:75-83.
———— 1958. Nameless streams: proposed new terms. *Cave Res. Grp Gt Br. Newsl.*, **72-77**:22-3.
———— 1963. Flow markings. *Cave Res. Grp Gt Br. Newsl.*, **87**:8-9.
Gregory, J. W. 1911. Constructive waterfalls. *Scott. geogr. Mag.*, **27**:537-46.
Groom, G. E. and Coleman, A. 1958. *The geomorphology and speleogenesis of the Dachstein Caves*. Cave Res. Grp Gt Br. Occ. Pub., **2**.
———— and Williams, V. H. 1965. The solution of limestone in South Wales. *Geogrl J.*, **131**:37-41.
Gross, M. G. 1964. Variations in the O^{18}/O^{16} and C^{13}/C^{12} ratios of diagenetically altered limestones in the Bermuda Islands. *J. Geol.*, **72**:170-94.
Grund, A. 1903. Die Karsthydrographie. *Geogr. Abh.*, **7**:3.
———— 1910a. Zur Frage des Grundwassers im Karste. *Mitt. geol. Ges. Wien*, **53**:606-17.
———— 1910b. *Das Karstphänomen*. Geol. Charakterbilder, **3**.
———— 1914. Der geographische Zyklus im Karst. *Z. Ges. Erdk. Berl.*, **1914**:621-40.
Halliday, W. R. 1957. *The Origin of the Limestone Caves of the Sierra Nevada of California*. Bull. Western Speleol. Surv. Misc. Ser., **3**.
———— 1960. Changing concepts of speleogenesis. *Bull. natn. speleol. Soc.*, **22**:23-9.
Hancock, P. L. 1968. Joints and faults: the morphological aspects of their origins. *Proc. Geol. Ass.*, **79**:146-51.
Harrison, J. V. 1930. The geology of some salt plugs in Laristan (Southern Persia). *Q.Jl geol. Soc. Lond.*, **86**:463-520.
Haserodt, K. 1969. Beobachtungen zur Karstdenudation am Kluftkarren in

glazialüberformten alpinen Bereichen. Pp. 123-38 in *Problems of the Karst Denudation*, ed. O. Štelcl. Brno.

Hem, J. D. 1959. *Study and Interpretation of the Chemical Characteristics of Natural Water*. U.S. geol. Surv. Wat.-Supply Pap., **1473**.

Hendy, C. H. 1969. Isotopic studies of speleothems. *New Zealand Spel. Soc. Bull.*, **4**:306-19.

Hodgkin, E. P. 1964. Rate of erosion of intertidal limestone. *Z.Geomorph.*, **8**:385-92.

Holland, H. D., Kirsipu, T. V., Huebner, J. S., and Oxburgh, U. M. 1964. On some aspects of the chemical evolution of cave waters. *J. Geol.*, **72**:36-67.

Holmes, J. W. and Colville, J. S. 1970a. Grassland hydrology in a karstic region of southern Australia. *Jnl Hydrol.*, **10**:38-58.

——— and ——— 1970b. Forest hydrology in a karstic region of southern Australia. *Jnl Hydrol.*, **10**:59-74.

Hooper, J. H. D. 1958. Bat erosion as a factor in cave formation. *Nature, Lond.*, **182**:1464.

Horberg, L. 1949. Geomorphic history of the Carlsbad Caverns area, New Mexico. *J. Geol.*, **57**:464-76.

Howard, A. D. 1963. The development of karst features. *Bull. natn. speleol. Soc.*, **25**:45-65.

Hundt, R. 1950. *Erdfalltektonik*. Halle.

Hunt, G. S. 1970. The origin and development of Mullamullang Cave N 37, Nullarbor Plain, Western Australia. *Helictite*, **8**:3-22.

Ineson, J. 1962. A hydrogeological study of the permeability of the chalk. *J. Instn. Water Engns.*, **16**:449-63.

Jennings, J. N. 1963. Geomorphology of the Dip Cave, Wee Jasper, New South Wales. *Helictite*, **1(3)**:43-58.

——— 1964. Geomorphology of Punchbowl and Signature Caves, Wee Jasper, New South Wales. *Helictite*, **2**:57-80.

——— 1966. Jirí V. Daneš and the Chillagoe Caves District. *Helictite*, **4**:83-7.

——— 1967a. Some karst areas of Australia. Pp. 256-92 in *Landform Studies from Australia and New Guinea*, ed. J. N. Jennings and J. A. Mabbutt. Canberra.

——— 1967b. Further remarks on the Big Hole, near Braidwood, New South Wales. *Helictite*, **6**:3-9.

——— 1967c. The surface and underground geomorphology. Pp. 13-31 in *Caves of the Nullarbor*, ed. J. R. Dunkley and T. M. L. Wigley. Sydney.

——— 1968. Syngenetic Karst in Australia. Pp. 41-110 in *Contributions to the Study of Karst*, P. W. Williams and J. N. Jennings. Aust. Nat.Univ.Dept.Geogr.Pub., **G/5**. Canberra.

——— 1969. Karst of the seasonally humid tropics in Australia. Pp. 149-58 in *Problems of the Karst Denudation*, ed. O. Štelcl. Brno.

——— and Bik, M. J. 1962. Karst morphology in Australian New Guinea. *Nature, Lond.*, **194**:1036-8.

——— and Sweeting, M. M. 1963. *The Limestone Ranges of the Fitzroy Basin, Western Australia*. Bonn. geogr. Abh., **32**.

Jordan, G. F. 1954. Large sink holes in the Straits of Florida. *Bull. Am. Ass. Petrol. Geol.*, **38**:1810-17.

Jordan, R. H. 1950. An interpretation of Floridian Karst. *J. Geol.*, **58**:261-8.

Katzer, F. von 1909. *Karst und Karsthydrographie.* Zur Kunde der Balkanhalbinsel, **8**. Sarajevo.

Kayser, K. 1934. Morphologische Studien in Westmontenegro II. *Z.Ges. Erdk. Berl.,* **1934**:26-49, 81-102.

Keller, W. D. 1957. *The principles of chemical weathering.* Columbia, Miss.

Kendall, P. F. and Wroot, H. E. 1924. *The Geology of Yorkshire.* Vienna.

King, L. 1951. The geology of the Makapan and other caves. *Trans. R. Soc. S.Afr.,* **33**:121-51.

King-Webster, W. A. and Kenny, J. S. 1958. Bat erosion as a factor in cave formation. *Nature, Lond.,* **181**:1813.

Kirkaldy, J. F. 1950. Solution of the chalk in the Mimms Valley, Herts. *Proc. Geol. Ass.,* **61**:219-24.

Klaer, W. von 1957. Karstkegel, Karst-Inselberg und Poljeboden am Beispiel des Jezeropoljes. *Petermanns geogr. Mitt.,* **101**:108-11.

Krebs, N. 1929. Ebenheiten und Inselberge im Karst. *Z. Ges. Erdk. Berl.,* **1929**:81-94.

Krejčí-Graf, K. 1935. Felsen aus Salz im Rumänien. *Natur Volk,* **65**:116-20.

Kukla, J. and Ložek, V. 1958. K problematice výzkumu jeskynnich výplni. *Cslký Kras,* **11**:19-59.

Kunský, J. 1958. *Karst et Grottes.* Trans. Heintz. Nr. 1399. Service d'Information Géologique du Bureau de Recherches Géologiques et Minières, Paris.

Kyrle, G. 1923. *Grundriss der Theoretische Speläologie.* Vienna.

Lange, A. L. 1962. Water level planes in caves. *Cave Notes,* **4**:12-16.

Lasserre, G. 1954. Notes sur le karst de Guadeloupe. *Erdkunde,* **8**:115.

La Valle, P. 1967. Some aspects of linear karst depression development in south central Kentucky. *Ann. Ass. Am. Geogr.,* **57**:49-71.

Lehmann, H. 1936. *Morphologische Studien auf Java.* Stuttgart.

———— 1953. Der tropische Kegelkarst in Westindien. *Verh. dt. Geogr. Tags.,* **29**:126-31.

———— 1959. Studien über Poljen in den venaziaschen Voralpen und im Hochapennin. *Erdkunde,* **12**:258-89.

———— 1960. La terminologie classique du karst sous l'aspect critique de la morphologie climatique moderne. *Revue Géogr. Lyon,* **35**:1-6.

————, Krommelbein, K. and Lötschert, W. 1956. Karstmorphologische, geologische und botanische Studien in der Sierra de los Organos auf Cuba. *Erdkunde,* **10**:185-204.

————, Roglić, J., Rathjens, C., Lasserre, G., Harrassowitz, H., Corbel, J., and Birot, P. 1954. Das Karstphänomen in den verschiedenen Klimazonen. *Erdkunde,* **8**:112-22.

Lehmann, O. 1927. Das Tote Gebirge als Hochkarst. *Mitt. geogr. Ges. Wien,* **70**:201-42.

———— 1932. *Die Hydrographie des Karstes.* Leipzig.

Leighton, M. W. and Pendexter, C. 1962. Carbonate rock types. *Mem. Am. Ass. Petrol. Geol.,* **1**:33-61.

Louis, H. 1956. Die Entstehung der Poljen und ihre Stellung im Taurus. *Erdkunde,* **10**:339-53.

Lowry, D. C. 1964. The development of Cocklebiddy Cave, Eucla Basin, Western Australia. *Helictite,* **3**:15-19.

———— 1967a. The origin of Easter Cave doline, Western Australia. *Aust. Geogr.,* **10**:300-2.

———— 1967b. Halite speleothems from the Nullarbor Plain, Western Australia. *Helictite,* **6**:14-20.

Lowry, D. C. 1969. The origin of small cavities in the limestone of the Bunda Plateau, Eucla Basin. *Geol. Surv. W. Aus. Ann. Rpt*, **1968**:34-7.

Ložek, V., Sekyra, J., Kukla, J., and Feyfar, O. 1956. Výskum Velke Jasovské Jeskyne. *Anthropozoikum*, **6**:193-282.

Lozinski, W. von 1907. Die Karsterscheinungen in Galazisch-Podolien. *Jber. geol. Bundesanst., Wien*, **1907**:683-726.

Marker, M. E. and Brook, G. A. 1970. *Echo Cave: a Tentative Quaternary Chronology for the Eastern Transvaal*. Dept. Geog. Env. Stud. Univ. Witwatersrand, Occ. Pap. **3**.

Martel, É. 1910. La théorie de la 'Grundwasser' et les eaux souterraines du karst. *Géographie*, **21**:126-30.

——— 1921. *Nouveau traité des eaux souterraines*. Paris.

Martin, J. 1965. Quelques types de dépressions karstiques du Moyen Atlas central. *Revue Géogr. maroc.*, **7**:95-106.

Matschinski, M. 1962. Sur le problème d'alignement de données apparemment dispersées. *C.r. hebd. Séanc.Acad.Sci.,Paris*, **254**:806-8.

Maucci, W. 1960. La speleogenesi nel Carso Triestino. *Boll. Soc. adriat. Sci. nat.*, **51**:233-54.

Melik, A. 1955. *Kraska Polja Slovenije v Pleistocenu*. Slovensk Akad. Znanosti in Umetnosti, 7, Inst. Geogr., 3.

Messines, J. 1948. Les éboulements dans les gypses. Inefficacité des travaux de reboisement. *C.r. hebd. Séanc.Acad.Sci.,Paris*, **226**:1295-6.

Miller, V. C. 1953. *A quantitative geomorphic study of drainage basin characteristics in the Clinch Mountain area, Virginia and Tennessee*. Dept. Geol. Columbia Univ. Tech. Rpt, 3, New York.

Mistardis, G. 1968. Investigations upon influences of sea level fluctuations on underground karstification in some coastal regions of south Greece. *Congr. 4ᵉ int. Spéléol.*, **3**:335-40.

Moneymaker, B. C. 1941. Subriver solution cavities in the Tennessee Valley. *J. Geol.*, **49**:74-86.

Monroe, W. H. 1966. Formation of tropical karst topography by limestone solution and reprecipitation. *Caribb. J. Sci.*, **6**:1-7.

——— 1968. The karst features of northern Puerto Rico. *Bull. natn. speleol. Soc.*, **30**:75-86.

——— 1970. *A glossary of karst terminology*. U.S. Geol. Surv. Water-supply Pap. 1899K.

Montoriol-Pous, J. 1951. Los processos clasticos hypogeos. *Rass. speleol. ital.*, **3**:119-29.

——— 1954. Resultado de nuevas observaciones sobre los processos clasticos hypogeos. *Rass. speleol. ital.*, **6**:103-14.

Moore, G. W. 1954. *The Origin of Helictites*. Occ. pap. natn. speleol. Soc., **1**.

——— 1960. Geology of Carlsbad Caverns, New Mexico. *Natn. Speleol. Soc. Guide Book*, **1**:10-18.

——— 1962. The growth of stalactites. *Bull. natn. speleol. Soc.*, **24**:95-106.

——— and Nicholas, G. 1964. *Speleology—the Study of Caves*. Boston.

Morawetz, S. 1967. Zur Frage der Karstebenheiten. *Z. Geomorph.*, **11**:1-13.

Morehouse, D. F. 1968. Cave development via the sulfuric acid reactions. *Bull. natn. speleol. Soc.*, **30**:1-10.

Myers, J. O. 1948. The formation of Yorkshire caves and potholes. *Trans. Cave Res. Grp Gt Br.*, **1(1)**:26-9.

Muxart, R., Stchouzkoy, T., and Franck, J. C. 1968. Observations hydro-karstologiques dans le bassin amont de la Seille (Jura). *Congr. 4ᵉ int. Spéléol.*, **3**:175-80.

Oertli, H. von 1953. Karbonathärte von Karstgewässern. *Stalactite*, **4(4)**: 8-18.

Oldfield, F. 1960. Studies in the Postglacial history of British vegetation: lowland Lonsdale. *New Phytol.*, **59**:192-217.

Ollier, C. D. 1963. The origin of limestone caves. *Helictite*, **1(2)**:8-12.

———— 1964. McEachern Cave, Nelson. *Vict. Nat.*, **81**:195-7.

———— 1969. *Volcanoes*. Canberra.

———— and Harrop, J. F. 1963. The caves of Mont Hoyo, Eastern Congo Republic. *Bull. natn. speleol. Soc.*, **25**:73-8.

———— and Holdsworth, D. K. 1968. Caves of Kiriwina, Trobriand Islands, Papua. *Helictite*, **6**:63-72.

———— and ———— 1969. Caves of Vakuta, Trobriand Islands, Papua. *Helictite*, **7**:50-61.

———— and ———— 1970. Some caves of Kitava, Trobriand Islands, Papua. *Helictite*, **8**:29-38.

———— and Tratman, E. K. 1969. Geomorphology of the caves. Ch. 4 in *The Caves of North-West Clare, Ireland*, ed. E. K. Tratman. Newton Abbot.

Ongley, E. D. 1968. An analysis of the meandering tendency of Serpentine Cave, N.S.W. *Jnl Hydrol.*, **6**:15-32.

Panoš, V. and Štelcl, O. 1968. Physiographic and geologic control in development of Cuban mogotes. *Z. Geomorph.*, **12**:117-65.

Pardé, M. 1965. *Influences de la perméabilité sur le régime des rivières*. Colloquium Geographicum, 7. Bonn.

Parker, G. G. 1964. Piping: a geomorphic agent in landform development of the drylands. *Int. Ass. Sci. Hydrol. Pub.*, **65**:103-13.

Penck, A. 1924. Das unterirdische Karstphänomen. Pp. 175-97 in *Zbornik Radova Posvećen Jovanu Cvijiću*, ed. P. Vujević. Belgrade.

Pengelly, W. 1864. The introduction of cavern accumulations. *Rep. Trans. Devon. Ass. Advmt. Sci.*, **3**:31-41.

Pevalek, I. 1935. Der Travertin und die Plitvice Seen. *Verh. int. Verein. theor. angew. Limnol.*, **7**:165-81.

Pfeffer, K-H. von 1967. Neue Beobachtungen im Kegelkarst von Jamaica. *Verh. dt. Geogr. Tags.*, **36**:345-58.

———— 1969. Charakter der Verwitterungsdecken im tropischen Kegelkarst und ihre Beziehung zum Formenschatz. *Geol. Rdsch.*, **58**:408-26.

Pigott, C. D. 1962. Soil formation and development on the Carboniferous Limestone of Derbyshire: parent materials. *J. Ecol.*, **50**:145-56.

Pinchemel, P. 1954. *Les Plaines de Craie du Nord-ouest du Bassin Parisien et du Sud-est du Bassin de Londres et leurs Bordures*. Paris.

Pitty, A. F. 1966. *An Approach to the Study of Karst Water: Illustrated by Results from Poole's Cavern, Buxton*. Univ. Hull Occ. Paper Geogr., **5**.

———— 1968. The scale and significance of solutional loss from the limestone tract of the Southern Pennines. *Proc. Geol. Ass.*, **79**:153-78.

Playford, P. D. and Lowry, D. C. 1966. *Devonian Reef Complexes of the Canning Basin, Western Australia*. Bull. Geol. Surv. W.Aust., **118**.

Pluhar, A. and Ford, D. C. (1970). Dolomite Karren of the Niagara escarpment, Ontario, Canada. *Z. Geomorph.*, **14**:392-410.

Pohl, E. R. 1955. *Vertical Shafts in Limestone Caves*. Occ. pap. natn. speleol. Soc., **2**.

———— and White, W. B. 1965. Sulfate minerals; their origin in the central Kentucky karst. *Am. miner.*, **50**:1461-5.

Price, H. J. 1959. Mechanics of jointing rocks. *Geol. Mag.*, **96**:149-67.

Priesnitz, K. von 1969. Über die Vergleichbarkeit von Lösungsformen auf Chlorid–, Sulfat– und Karbonatgestein–Überlegungen zu Fragen der Nomenklatur und Methodik der Karstmorphologie. *Geol. Rdsch.*, **58**:427-38.

Prinz, W. 1908. *Les cristallisations des grottes de Belgique.* Brussels.

Rathjens, C. 1954. Zur Frage der Karstrandebene im dinarischen Karst. *Erdkunde*, **8**:114-15.

———— 1960. Beobachtungen an hochgelegenen Poljen im südlichen dinarischen Karst. *Z. Geomorph.*, **4**:141-51.

Reams, M. W. 1968. Cave sediments and the geomorphic history of the Ozarks. Ph.D. thesis, Washington University, St Louis, Miss.

Reid, C. 1887. On the origin of dry chalk valleys and of coombe rock. *Q. Jl geol. Soc. Lond.*, **43**:364-73.

Renault, P. 1958. Éléments de spéléomorphologie karstique. *Ann. Spéléol.*, **13**:23-48.

———— 1959. Le karst Kouilou (Moyen Congo, Gabon). *Revue Géogr. Lyon*, **1959**:305-14.

———— 1967-8. Contribution à l'étude des actions mécaniques et sédimentologiques dans la spéléogenèse. *Ann. Spéléol.*, **22**:5-21, 209-67; **23**:259-307.

Rhoades, R. and Sinacori, M. N. 1941. Pattern of groundwater flow and solution. *J. Geol.*, **49**:785-94.

Riedl, H. R. 1961. Gedanken zur methodischer Intensivierung der speläogenetischen Forschung. *Petermanns geogr. Mitt.*, **105**:264-8.

Roglić, J. 1939. Morphologie der Poljen von Kupres und Vukovsko. *Z. Ges. Erdk. Berl.*, **1939**:299-316.

———— 1940. Geomorphologische Studien von Duvanjsko Polje (Polje von Duvno) in Bosnien. *Mitt. geogr. Ges. Wien*, **83**:152-76.

———— 1954. Korrosive Ebenen im dinarischen Karst. *Erdkunde*, **8**:113-14.

———— 1964a. Karst valleys in the Dinaric Karst. *Erdkunde*, **18**:113-16.

———— 1964b. Les poljes du karst dinarique et les modifications climatiques du Quaternaire. *Rev. belge Géogr.*, **88**:105-25.

Roques, H. 1964. Contribution à l'étude statique et cinétique des systèmes gaz carbonique-eau-carbonate. *Ann. Spéléol.*, **19**:255-484.

———— 1969. A review of present-day problems in the physical chemistry of carbonates in solution. *Trans. Cave Res. Grp Gt Br.*, **11**:139-64.

Sanders, E. M. 1921. The cycle of erosion in a karst region (after Cvijić). *Geogr. Rev.*, **11**:593-604.

Sawicki, L. R. von 1909. Ein Beitrag zum geographischen Zyklus im Karst. *Geogr. Ztschr.*, **15**:185-204, 259-81.

Schmid, E. 1958. *Höhlenforschung und Sedimentanalyse. Ein Beitrag zur Datierung des Alpinen Paläolithikums.* Institut für Ur- und Frühgeschichte der Schweiz. Schriften 13, Basel.

Schoemaker, R. P. 1948. A review of rock pressure problems. *Mining Technology*, **12** (TP2495):1-14.

Shaw, T. R. and Tratman, E. K. 1969. Mainly historical. Pp. 15-32 in *The Caves of North-West Clare, Ireland*, ed. E. K. Tratman. Newton Abbot.

Siffre, A. and Siffre, M. 1961. Le façonnement des alluvions karstiques. *Ann. Spéléol.*, **16**:73-80.

Siffre, M. 1959. Ecoulement nappo-laminaire et morphologie souterraine. *Stalactite*, **4**:49-55.

Silar, J. 1965. Development of tower karst of China and North Vietnam. *Bull. natn. speleol. Soc.*, **27**:35-46.

Smith, D. I. 1965. Some aspects of limestone solution in the Bristol region. *Geogrl J.*, **131**:44-9.

———— 1969. The solution erosion of limestone in an arctic morphogenetic region. Pp. 99-110 in *Problems of the Karst Denudation*, ed. O. Štelcl. Brno.

————, High, C., and Nicholson, F. H. 1969. Limestone solution and the caves. Pp. 96-123 in *The Caves of North-West Clare, Ireland*, ed. E. K. Tratman. Newton Abbot.

———— and Mead, D. G. 1962. The solution of limestone. *Proc. speleol. Soc. Univ. Bristol*, **9**:188-211.

Smyk, B. and Drzal, M. 1964. Research on the influence of microorganisms on the development of karst phenomena. *Geographia Polonica*, **2**: 57-60.

Sparks, B. W. 1961. *Geomorphology*. London.

———— and Lewis, W. V. 1957. Escarpment dry valleys near Pegsdon, Hertfordshire. *Proc. Geol. Ass.*, **68**:26-38.

Štelcl, O., Vlček, V., and Piše, J. 1969. Limestone solution intensity in the Moravian karst. Pp. 71-86 in *Problems of the Karst Denudation*, ed. O. Štelcl. Brno.

Stringfield, V. T. and LeGrand, H. E. 1969a. Hydrology of carbonate rock terranes—a review. *Jnl Hydrol.*, **8**:349-413.

———— and ———— 1969b. Relation of sea water to fresh water in carbonate rocks in coastal areas, with special reference to Florida, U.S.A. and Cephalonia (Kephallinia), Greece. *Jnl Hydrol.*, **9**: 387-404.

Sunartadirdja, M. A. and Lehmann, H. 1960. Der tropische Karst von Maros und Nord-Bone in SW-Celebes (Sulawesi). *Z.Geomorph.-Suppl.*, **2**:49-65.

Sweeting, M. M. 1950. Erosion cycles and limestone caverns in the Ingleborough District of Yorkshire. *Geogrl J.*, **115**:63-78.

———— 1953. The enclosed depression of Carran, County Clare. *Ir. Geogr.*, **2**:218-24.

———— 1955. The landforms of North-West County Clare, Ireland. *Trans. Inst. Br. Geogr.*, **21**:33-49.

———— 1958. The karstlands of Jamaica. *Geogrl J.*, **124**:184-99.

———— 1965. Denudation in limestone regions, I Introduction. *Geogrl J.*, **131**:34-7.

———— 1966. The weathering of limestones. With particular reference to the Carboniferous limestones of northern England. Pp. 177-210 in *Essays in Geomorphology*, ed. G. H. Dury. London.

———— 1968. Karstic morphology. In The University of Edinburgh British-Honduras-Yucatan Expedition. *Geogrl J.*, **134**:49-54.

Swinnerton, A. C. 1929. The caves of Bermuda. *Geol. Mag.*, **66**:79-84.

———— 1932. Origin of limestone caverns. *Bull. geol. Soc. Am.*, **43**:662-93.

Symposium on Cave Surveying 1970. *Trans. Cave Res. Grp Gt Br.*, **12(3)**: 1-245.

Symposium on Cave Hydrology and Water Tracing. *Trans. Cave Res. Grp Gt Br.*, **10(2)**:1-125.

Thomas, T. M. 1954. Swallow holes on the Millstone Grit and Carboniferous Limestone of the South Wales Coalfield. *Geogrl J.*, **120**: 468-75.

———— 1963. Solution subsidence in south-east Carmarthenshire and south-west Breconshire. *Trans. Inst. Br. Geogr.*, **33**:45-60.

Thorp, J. 1934. The asymmetry of the Pepino Hills of Puerto Rico in relation to the Trade Winds. *J. Geol.*, **42**:537-45.

Thrailkill, J. V. 1960. Origin and development of Fulford Cave, Colorado. *Bull. natn. speleol. Soc.*, **22**:54-65.

———— 1968. Chemical and hydrologic factors in the excavation of limestone caves. *Bull. geol. Soc. Am.*, **79**:19-45.

Tratman, E. K. 1969. The rate of solution of carboniferous limestone in the caves of north west Clare, Ireland. Pp. 87-98 in *Problems of the Karst Denudation*, ed. O. Štelcl. Brno.

Tricart, J. 1955. Modelé karstique et modelé périglaciaire dans les Causses. *Revue géomorph. dyn.*, **6**:193-201.

———— 1968. Notes géomorphologiques sur la karstification en Barbade (Antilles). *Mém. Docums. Cent.Docum. cartogr. géogr.*, **4**:329-34.

———— and Silva, T. C. da 1960. Un exemple d'évolution karstique en milieu tropical sec: Le morne de Bom Jesus da Lapa (Bahia, Brésil). *Z. Geomorph.*, **4**:29-42.

Trimmel, H. 1950. Beobachtungen zur Frage der Raumbildung in der Badlhöhle im Mittelsteirischen Karst. *Mitt. geogr. Ges. Wien*, **92**:26-30.

———— 1951. Morphologische und genetische Studien in der Salzofenhöhle. *Höhle*, **2**:2-7.

———— (ed.) 1965. *Speläologisches Fachwörterbuch*. Congr. 3ᵉ int. Spéléol., C.

———— 1968. *Höhlenkunde*. Brunswick.

Trombe, F. 1952. *Traité de Spéléologie*. Paris.

Twidale, C. R. 1971. *Structural Landforms*. Canberra.

Varnes, D. J. 1958. Landslide types and processes. Pp. 20-47 in *Landslides and Engineering Practice*, ed. E. B. Eckel. Highway Research Board Spec. Ref., 29.

Vaumas, E. de 1968. Phénomènes karstiques en Méditerranée orientale. *Mém. Docums. Cent.Docum. cartogr. géogr.*, **4**:193-282.

Veeh, H. H. and Chappell, J. 1970. Astronomical theory of climatic change: support from New Guinea. *Science, N.Y.*, **167**:862-5.

Verstappen, H. Th. 1960a. On the geomorphology of raised coral reefs and its tectonic significance. *Z. Geomorph.*, **4**:1-28.

———— 1960b. Some observations on karst development in the Malay Archipelago. *J. trop. Geogr.*, **14**:1-10.

———— 1964. Karst morphology of the Star Mountains (central New Guinea) and its relation to lithology and climate. *Z. Geomorph.*, **8**:40-9.

Wagner, G. 1935. Vom Salz des Toten Meeres. *Natur Volk*, **65**:108-16.

Wall, J. R. D. and Wilford, G. E. 1966. Two small-scale solution features of limestone outcrops in Sarawak, Malaysia. *Z. Geomorph.*, **10**:90-4.

Warwick, G. T. 1950. The reef limestone caves of the Dove and Manifold Valleys. *Cave Res. Grp Gt Br. Newsl.*, **31**:2-6.

———— 1953. Polycyclic swallow holes in the Manifold valley, Staffordshire, England. *Congr. 1ᵉʳ int. Spéléol.*, **2**:59-68.

———— 1958. The characteristics and development of limestone regions in the British Isles with special reference to England and Wales. *Congr. 2ᵉ int. Spéléol.*, **1**:79-105.

———— 1960. The effect of knick-point recession on the watertable and associated features in limestone regions, with special reference to England and Wales. *Z. Geomorph.-Suppl.*, **2**:92-7.

Warwick, G. T. 1962. Cave formations and deposits. Pp. 83-119 in *British Caving*, ed. C. H. D. Cullingford. London.

—— 1964. Dry valleys in the southern Pennines, England. *Erdkunde*, **18**:116-23.

—— 1968. Some primitive features in British caves. *Congr. 4ᵉ int. Spéléol.*, **3**:239-52.

Weyl, P. K. 1958. The solution kinetics of calcite. *J. Geol.*, **66**:163-76.

White, E. L. and Reich, B. M. 1970. Behaviour of annual floods in limestone basins in Pennsylvania. *Jnl Hydrol.*, **10**:193-8.

White, W. B. 1960. Termination of passages in Appalachian caves as evidence for a shallow phreatic origin. *Bull. natn. speleol. Soc.*, **22**:43-53.

—— 1969. Conceptual models for carbonate aquifers. *Ground Water*, **7**:15-21.

——, Jefferson, G. L., and Haman, J. F. 1966. Quartzite karst in southeastern Venezuela. *Int. J. Speleol.*, **2**:309-14.

Wilford, G. E. 1964. *The geology of Sarawak and Sabah Caves*. Bull. geol. Surv. Borneo Region, Malaysia, **6**.

—— and Wall, J. R. D. 1965. Karst topography in Sarawak. *J. trop. Geogr.*, **21**:44-70.

Williams, P. W. 1963. An initial estimate of the speed of limestone solution in County Clare. *Ir. Geogr.*, **4**:432-41.

—— 1966a. Limestone pavements with special reference to Western Ireland. *Trans. Inst. Br. Geogr.*, **40**:155-72.

—— 1966b. Morphometric analysis of temperate karst landforms. *Ir. Speleol.*, **1**:23-31.

—— 1968. An evaluation of the rate and distribution of limestone solution in the River Fergus basin, western Ireland. Pp. 1-40 in *Contributions to the Study of Karst*. Aust. National Univ. Dept Geogr. Pub. **G/5**. Canberra.

—— 1969. The geomorphic effects of ground water. Pp. 269-84 in *Water, Earth and Man*, ed. R. J. Chorley. London.

—— 1970. Limestone morphology in Ireland. Ch. 7 in *Irish Geographical Studies*, ed. R. Glosscock and N. Stephens. Belfast.

Wissmann, H. von 1954. Der Karst der humiden heissen und sommerheissen Gebiete Ostasiens. *Erdkunde*, **8**:122-30.

—— 1957. Karsterscheinungen in Hadramaut. Ein Beitrag zur Morphologie der semiariden und ariden Tropen. *Petermanns Mitt. Erg.*, **262**:259-68.

Woodward, H. P. 1936. Natural Bridge and Natural Tunnel, Virginia. *J. Geol.*, **44**:604-16.

Wright, F. J. 1936. The Natural Bridge of Virginia. *Va. Geol. Surv. Bull.*, **46G**:50-78.

Würm, A. 1953. Der Salzberg bei Djelfa im Sahara-Atlas. *Natur Volk*, **83**:141-7.

Zötl, J. 1957. Neue Ergebnisse der Karsthydrologie. *Erdkunde*, **11**:107-17.

—— 1965. Tasks and results of karst hydrology. Pp. 141-5 in *Problems of the Speleological Research*, ed. O. Štelcl. Prague.

Zotov, V. D. 1941. Pot-holing of limestone by development of solution cups. *J. Geomorph.*, **4**:71-3.

INDEX

N.S.W. = New South Wales; Vic. = Victoria; Qld = Queensland; Tas. = Tasmania; S.A. = South Australia; W.A. = Western Australia; N.Z. = New Zealand; N.G. = Territory of Papua and New Guinea; Craven = Craven, Yorkshire; Peak = Peak District, Derbyshire; Mendip = Mendip Hills, Somerset; Clare = County Clare, Ireland.

The names of the individual states and provinces of the United States of America and the Federal Republic of Yugoslavia are employed alone.

Bold face indicates page reference to figures, plates, and tables.